# Antifungal and Antiparasitic Drug Delivery

# Antifungal and Antiparasitic Drug Delivery

Special Issue Editors

**Juan José Torrado Durán**
**Dolores R. Serrano**
**Javier Capilla**

MDPI • Basel • Beijing • Wuhan • Barcelona • Belgrade • Manchester • Tokyo • Cluj • Tianjin

MDPI

*Special Issue Editors*
Juan José Torrado Durán
Complutense University of Madrid
Spain

Dolores R. Serrano
Complutense University of Madrid
Spain

Javier Capilla
Universitat Rovira i Virgili and Institut d'investigatió Sanitaria Pere Virgili (IISPV)
Spain

*Editorial Office*
MDPI
St. Alban-Anlage 66
4052 Basel, Switzerland

This is a reprint of articles from the Special Issue published online in the open access journal *Pharmaceutics* (ISSN 1999-4923) (available at: https://www.mdpi.com/journal/pharmaceutics/special_issues/antifungal_DD).

For citation purposes, cite each article independently as indicated on the article page online and as indicated below:

LastName, A.A.; LastName, B.B.; LastName, C.C. Article Title. *Journal Name* **Year**, *Article Number*, Page Range.

**ISBN 978-3-03936-306-3 (Pbk)**
**ISBN 978-3-03936-307-0 (PDF)**

# Contents

# About the Special Issue Editors

**Juan José Torrado Durán**, Pharmaceutics and Food Technology, School of Pharmacy, Complutense University of Madrid, Madrid, Spain. He is author of about 100 articles related to Pharmaceutical Technology and Drug Delivery Systems. Interests: conventional pharmaceutical dosage forms (tablets, capsules, semisolid and liquid formulations); new controlled release systems (pellets, nanoparticles, microcapsules, microspheres and liposomes) including production and quality control.

**Dolores R. Serrano**, Pharmaceutics and Food Technology, School of Pharmacy, Complutense University of Madrid, Madrid, Spain. She is author of about 60 articles related to Pharmaceutical Technology and Drug Delivery Systems. Interests: development and optimization of novel drug delivery systems of poorly water soluble drugs with the aim of increasing their bioavailability and their effectiveness with special focus on antifungals and antiparasitic drugs, as well as 3D printing of medicines.

**Javier Capilla**, Unitat de Microbiologia, Facultat de Medicina i Ciènces de la Salut, Universitat Rovira i Virgili and Institut d'investigatió Sanitaria Pere Virgili (IISPV), Reus, Spain. He is author of about 70 articles related to fungal diseases and their treatments. Interests: the study and development of antifungal therapies against opportunistic fungal infections as well as virulence factors of fungi causing human infections.

*pharmaceutics*

MDPI

*Editorial*

# Antifungal and Antiparasitic Drug Delivery

**Juan José Torrado [1,*], Dolores R. Serrano [1,*] and Javier Capilla [2]**

1 Departament of Pharmaceutics and Food Technology, School of Pharmacy, Complutense University of Madrid, 28040 Madrid, Spain

2 Unitat de Microbiologia, Facultat de Medicina i Ciències de la Salut, Universitat Rovira i Virgili and Institut d'Investigació Sanitària Pere Virgili (IISPV), 43201 Reus, Spain; javier.capilla@urv.cat

* Correspondence: torrado1@farm.ucm.es(J.J.T.); drserran@ucm.es (D.R.S.)

Received: 25 March 2020; Accepted: 1 April 2020; Published: 4 April 2020

**Abstract:** Fungal and parasitic diseases affect more than a billion people across the globe, one-sixth of the world's population, mostly located in developing countries. The lack of effective and safer treatments combined with a deficient diagnosis lead to serious chronic illness or even death. There is a mismatch between the rate of drug resistance and the development of new medicines. Formulation of antifungal and antiparasitic drugs adapted to different administration routes is challenging, bearing in mind their poor water solubility, which limits their bioavailability and efficacy. Hence, there is an unmet clinical need to develop vaccines and novel formulations and drug delivery strategies that can improve the bioavailability and therapeutic effect by enhancing their dissolution, increasing their chemical potency, stabilising the drug and targeting high concentration of drug to the infection sites. This Editorial regards the ten research contributions presented in the Special Issue "Antifungal and Antiparasitic Drug Delivery".

**Keywords:** liposomes; transferosomes; nanoparticles; emulsions; candidiasis; aspergillosis; azoles; amphotericin B; combined therapy; quality by design; leishmaniasis; malaria; trypanosomiasis

---

In order to obtain new antifungal and antiparasitic drug delivery systems, scientists of different disciplines have to collaborate in coordinated research teams. This volume includes ten papers, five of them about antifungal formulations and other five related to antiparasitic formulations. Amongst fungal infections, candidiasis has received special attention due to its world prevalence, as well as leishmaniasis as a parasitic disease. Interestingly, an old molecule, amphotericin B, is the active component studied in six out of the ten papers. Other active components also studied are butenafine, praziquantel, fluconazole, meglumine antimoniate and the enolase-base vaccine.

The administration route plays a key role in the development of novel antifungal and antiparasitic formulations. In this issue, a special focus on oral, parenteral and topical formulations is highlighted. Dosage forms are obviously related to the administration route. For example, suspensions, solutions and tablets are developed for oral administration while semisolid gels and wound patches are fabricated for topical application. Different parenteral administration routes are covered, such as subcutaneous, intravenous and at the bone cavity.

The originality of the new formulations proposed are usually based on the selection of already approved excipients along with the active components, such as Montanide™ Petgel A as vaccine adjuvant [1], poly(vinyl alcohol) [2], Poloxamer 407™ [3], a combination of Capryol 90™, Peceol™ and Labrasol™ [4], dextran and maltodextrin [5], a combination of modified chitosan nanoparticles with a standardized extract of cultured *Lentinula edodes* mycelia (AHCC™) [6], ground calcium carbonate [7], poly (D,L-lactide-*co*-glycolide) 50:50 [8] and Sepigel 305™ [9]. Only in one paper [10], authors have synthesized a new material based on copolymers of poly(ethyleneglycol) and poly(ε-caprolactone) conjugated with retinol as drug vehicle. Moreover, the characterization of the new formulations is described in detail in the ten papers with a special focus on toxicity and efficacy studies required in order to bring to the market these formulations. This Special Issue is an update on novel drug delivery strategies of antifungal and antiparasitic drugs to treat both topical and systemic infections. A brief description of the ten research papers included in the issue is described below.

Tellez-Martínez et al. propose a new vaccine based on recombinant enolase-Montanide™ PetGel A against virulent fungus *Sporothrix schenckii*. The incorporation of Montanide™ PetGel A as adjuvant was able to induce specific Th1 response and protective immunity against the fungal in Balb/c mice [1]. Interestingly, the virulence of *S. schenckii* was enhanced by toluene exposure. Toluene is an example of environmental contaminant. In this work, authors proved that the combination of some environmental contaminants can enhance the virulence of pathogen agents. Effective vaccines are an important pharmacological tool to protect us against this type of severe infections.

The work of Alexandrino-Junior et al. is a clear example of the potential pharmacological effect of new formulations of old drugs. Amphotericin B was formulated on a poly(vinyl-alcohol) hydrogel as a new topical formulation for the treatment of cutaneous leishmaniasis [2]. Although topical treatment of cutaneous diseases seems to be an ideal approach, conventional topical formulations of amphotericin possess low activity on cutaneous leishmaniasis due to permeability issues. Nevertheless, these new hydrogels developed in this work have exhibited, in vitro, a promising antiparasitic activity against *Leishmania* parasites and also against some fungal infections.

Sosa et al. performed an interesting study whose aim was the development and evaluation of a topical formulation of amphotericin B for the treatment of dermal and vaginal candidiasis [3]. Poloxamer 407™ was selected as excipient based on its thermoreversible properties. This excipient is liquid at low temperatures (4–5 °C) but turns into a semisolid gel above 32 °C. A thermoreversible gel containing amphotericin was developed and evaluated. Ex vivo permeation studies on human skin and pig vaginal mucosa showed that no permeation was observed. In vitro, antifungal activity studies against *Candida* spp showed that this formulation was more efficient than free amphotericin. Moreover, the amount of amphotericin remaining on the skin and vaginal mucosa was high enough to obtain antifungal activity.

Bezerra-Sousa et al. described the preparation of an oral nanomedicine of butenafine for visceral leishmaniasis [4]. Butenafine is currently used as a topical antifungal drug with low oral bioavailability. In this work, the low solubility of butenafine was increased by preparation of optimized self-nanoemulsifying drug delivery systems which have proved in vitro to be effective against promastigotes and amastigotes of *Leishmania infantum*. Moreover, these promising systems were then transformed by spray-drying into a solid dosage form of butenafine. Development of solid oral nanomedicines enables the non-invasive and safe drug administration, being a cost-effective and readily scalable repurposed medicine for visceral leishmaniasis.

Serrano et al.'s work focused on the design of fast-dissolving orodispersible films of amphotericin B for oropharyngeal candidiasis [5]. Amphotericin B is a low water soluble antifungal drug. A quality-by-design study was applied to select the best combination of GRAS excipients. A fast disintegration film with quick amphotericin release in artificial saliva and high in vitro efficacy against several *Candida* spp. was obtained.

Pérez-Cantero et al. carry out an interesting study related to the increased prophylactic efficacy of parenteral and oral amphotericin B treatments against aspergillosis when combined with standardized extract of cultured *Lentinula edodes* mycelia (AHCC™) [6]. Amphotericin was encapsulated in modified chitosan-nanoparticles suitable for oral administration. The addition of AHCC™ significantly improved the efficacy of both oral and parenteral treatments in a mice model of experimental aspergillosis. Moreover, the weight loss of treated animals was lower when AHCC™ was administered, suggesting a protective effect of the extract. In relation to the control group, treated animals showed stimulation of the Th1 immune response, which can explain the improvement of its efficacy.

The work of Borrego-Sánchez et al. focused on the increase of solubility and dissolution rate of praziquantel [7]. Praziquantel is also a poorly water-soluble antiparasitic drug, highly effective against schistosomiasis. Ground calcium carbonate is a cheap, hydrophilic porous carrier that was combined with praziquantel by using two easily scalable processes: physical mixture or solid dispersions. An in vitro dissolution test proved that solid dispersions increase drug solubility and dissolution rate. In vitro cytotoxicity studies against HTC116 cells showed that the praziquantel solid dispersions are safe.

Hsu et al. studied how amphotericin B and fluconazole can be incorporated into resorbable beads [8]. These beads are made of biodegradable Poly(D,L-lactide-*co*-glycolide) (50:50) and they were fabricated using a compression-molding method. The beads were evaluated, showing that the in vitro release of the fluconazole beads was better than the one obtained from amphotericin B beads. The in vivo assay in rabbits showed a sustained antifungal activity of fuconazole for more than 49 days, and thus, was suitable for the treatment of bone infections.

Berenguer et al. developed and characterized a semi-solid gel dosage form of meglumine antimoniate for the topical treatment of cutaneous leishmaniasis [9]. The gel is easy to prepare and its main excipient is Sepigel 305™. It was stable for over 6 months. The pH and rheological characteristics were suitable for topical application. Ex vivo permeation studies in human skin show low permeation and high retention in the skin layer, so low systemic toxicity and enhanced local activity can be expected from this formulation. Low toxicity and good tolerance were observed in keratinocyte cell lines and human volunteers, respectively. In vitro anti-leishmanial activity of the gel showed a reduction of the $IC_{50}$ compared to the reference solution. This new formulation could be a promising alternative for topical treatment of cutaneous leishmaniasis.

Rodriguez et al. described the development of amphotericin B micellar formulations based on copolymers of poly(ethyleneglycol) and poly(ε-caprolactone) conjugated with retinol [10]. Biodegradable and biocompatible polymers were initially synthesized and then conjugated with retinol. These micellar formulations were less haemolytic than Fungizone™. Furthermore, the antifungal activity of amphotericin incorporated in these new formulations showed a reduction of the MIC of up to eight-fold compared with reference Fungizone™. The low toxicity and high in vitro antifungal activity of these formulations make them good candidates for future in vivo experiments.

**Conflicts of Interest:** The authors declare no conflict of interest.

## References

1. Téllez-Martínez, D.; Portuondo, D.L.; Loesch, M.L.; Batista-Duharte, A.; Carlos, I.Z. A recombinant enolase-Montanide™ Petgel A vaccine promotes a protective Th1 immune response against a highly virulent *Sporothrix schenckii* by toluene exposure. *Pharmaceutics* **2019**, *11*, 144. [CrossRef] [PubMed]
2. Alexandrino-Junior, F.; Holanda e Silva, K.G.; Freire, M.C.L.C.; Lione, V.O.F.; Azevedo Cardoso, E.; Marcelino, H.R.; Genre, J.; Gomes de Oliveira, A.; Egito, E.S.T. A functional wound dressing as a potential treatment for cutaneous leishmaniasis. *Pharmaceutics* **2019**, *11*, 200. [CrossRef] [PubMed]
3. Sosa, L.; Calpena, A.C.; Silva-Abreu, M.; Espinoza, L.C.; Rincón, M.; Bozal, N.; Domenech, O.; Rodríguez-Lagunas, M.J.; Clares, B. Thermoreversible gel loaded amphotericin B for the treatment of dermal and vaginal candidiasis. *Pharmaceutics* **2019**, *11*, 312. [CrossRef] [PubMed]
4. Becerra-Sousa, A.; Fernández-García, R.; Rodrígues, G.F.; Bolás-Fernández, F.; Laurenti, M.D.; Passero, L.F.; Lalatsa, A.; Serrano, D.R. Repurposing butenafine as an oral nanomedicine for visceral leishmaniasis. *Pharmaceutics* **2019**, *11*, 353. [CrossRef] [PubMed]
5. Serrano, D.R.; Fernández-García, R.; Mele, M.; Healy, A.M.; Lalatsa, A. Designing fast-dissolving orodispersible films of amphotericin B for oropharyngeal candidiasis. *Pharmaceutics* **2019**, *11*, 369. [CrossRef] [PubMed]
6. Pérez-Cantero, A.; Serrano, D.R.; Navarro-Rodríguez, P.; Schätzlein, A.G.; Uchegbu, I.F.; Torrado, J.J.; Capilla, J. Increased efficacy of oral fixed-dose combination of amphotericin B and AHCC® natural adjuvant against aspergillosis. *Pharmaceutics* **2019**, *11*, 456. [CrossRef] [PubMed]
7. Borrego-Sánchez, A.; Sánchez-Espejo, R.; Albertini, B.; Passerini, N.; Cerezo, P.; Viseras, C.; Sainz-Díaz, C.I. Ground calcium carbonate as a low cost and biosafety excipient for solubility and dissolution improvement of praziquantel. *Pharmaceutics* **2019**, *11*, 533. [CrossRef] [PubMed]

8. Hsu, Y.-H.; Chen, H.-Y.; Chen, J.-C.; Yu, Y.-H.; Chou, Y.-C.; Ueng, S.W.-N.; Liu, S.-J. Resorbable beads provide extended release of antifungal medication: In vitro and in vivo analyses. *Pharmaceutics* **2019**, *11*, 550. [CrossRef] [PubMed]
9. Berenguer, D.; Sosa, L.; Alcover, M.; Sessa, M.; Halbaut, L.; Guillén, C.; Fisa, R.; Calpena-Campmany, A.C.; Riera, C. Development and characterization of a semi-solid dosage form of meglumine antimoniate for topical treatment of cutaneous leishmaniasis. *Pharmaceutics* **2019**, *11*, 613. [CrossRef] [PubMed]
10. Rodríguez, Y.J.; Quejada, L.F.; Villamil, J.C.; Baena, Y.; Parra-Giraldo, C.-M.; Perez, L.D. Development of amphotericin B micellar formulations based on copolymers of poly(ethylene glycol) and poly(ε-caprolactone) conjugated with retinol. *Pharmaceutics* **2020**, *12*, 196. [CrossRef]

*pharmaceutics*

MDPI

*Article*

# A Recombinant Enolase-Montanide™ PetGel A Vaccine Promotes a Protective Th1 Immune Response against a Highly Virulent *Sporothrix schenckii* by Toluene Exposure

**Damiana Téllez-Martínez, Deivys Leandro Portuondo, Maria Luiza Loesch, Alexander Batista-Duharte * and Iracilda Zeppone Carlos ***

Department of Clinical Analysis, School of Pharmaceutical Sciences, São Paulo State University (UNESP), Araraquara 14800-903, SP, Brazil; damianatellezm@gmail.com (D.T.-M.); deivysleandro@gmail.com (D.L.P.); ma_luizaloesch@hotmail.com (M.L.L.)

* Correspondence: batistaduhartea@gmail.com or batistaduhartea@fcfar.unesp.br (A.B.-D.); carlosiz@fcfar.unesp.br (I.Z.C.)

Received: 25 February 2019; Accepted: 21 March 2019; Published: 25 March 2019

**Abstract:** The effect of vaccination in fungal strains that suffered changes in their virulence by exposure to environmental contaminants is largely known. Growing reports of resistance to antifungal drugs and the emergence of new highly virulent strains, possibly acquired in the environment, prompt the design of new vaccines able to prevent and combat emerging mycotic diseases. In this study, we evaluated the protective capacity of an enolase-based vaccine and Montanide PetGel A (PGA) as an adjuvant against *S. schenckii* with increased virulence by exposure to toluene. The adjuvanted vaccine induced a strong specific Th1 response and protective immunity against a challenge with either wildtype or toluene-adapted *S. schenckii* in Balb/c mice. This study highlights the role of the adjuvant PGA driving the quality of the anti-sporothrix immunity and the key component in the vaccine efficacy.

**Keywords:** vaccine; adjuvants; *Sporothrix schenckii*; toluene; virulence; enolase; Montanide PetGel A

---

## 1. Introduction

Sporotrichosis is an emergent subcutaneous mycosis in tropical and subtropical regions, caused by several pathogenic species of the genus *Sporothrix*; that include *Sporothrix brasiliensis*, *S. schenckii sensu stricto*, *Sporothrix globosa*, and *Sporothrix luriei* [1]. Classically, infection is acquired after traumatic inoculation of contaminated soil, plants, and organic matter into skin or mucosa or, more rarely, by inhalation of conidia. Over the last years, cat–human zoonotic transmission of sporotrichosis caused by *S. brasiliensis* has become a health problem in Brazil [2]. The disease can manifest as fixed cutaneous and regional lymphocutaneous forms in immunocompetent individuals, and disseminated forms, mainly reported in immunocompromised patients [3].

Ecological determinants of the genus *Sporothrix* remain poorly understood [4,5]. However, experimental evidence suggests that environmental contaminants can modify the fungal virulence by reducing the host immunity [6] or modifying the fungal biology [4,7]. Previous studies showed that fungal exposure to toluene, a common soil contaminant that shares the same environmental niche of *S. schenckii*, is able to increase the *S. schenckii* virulence [7]. Ongoing studies are evaluating the role of chemical contamination and other environmental factors in sporotrichosis outbreaks.

Conventional treatment of sporotrichosis requires long periods of antifungal drug administration often accompanied by adverse effects and fungal resistance, principally during the treatment of disseminated sporotrichosis [8]. These problems have stimulated the search for new strategies for sporotrichosis management, including anti-sporothrix vaccination that has been proposed as a feasible way for both therapeutic and prophylactic purposes [9,10]. However, the development of antifungal vaccines has not been as successful as antiviral and antibacterial vaccines due to, among other things, a general under-appreciation for the impact of fungal diseases and the high cost of preclinical and clinical studies [11]. Another challenge has been the use of immunological adjuvants with an adequate safety and efficacy profile [12]. Aluminium-derived adjuvants have been used in human and veterinary vaccines for more than eight decades in licensed vaccines [13,14]. However, they have not been successful in preventing intracellular infection due to a weak capacity to induce cell-mediated immunity when used along with small immunogenic antigens [13]. Moreover, there are reports of tumors in the inoculation site in felines vaccinated with alum-based vaccines and a possible causal association between chronic inflammation induced by alum and these tumors has been suggested [15].

Current advances in the understanding of antifungal immune response support rational use of more effective adjuvants, such as pattern recognition receptors (PRR) agonists, inhibitors of regulatory T cells, and others [12,16,17], to achieve effective immune responses against specific fungi.

For several decades, biodegradable natural and synthetic polymers have been used for antigen delivery and as immunological adjuvants. Due to their biocompatibility, biodegradability, easy production and low toxicity polymers are attractive candidates for substituting conventional adjuvants [18]. MontanideTM PetGel A (PGA), is a polymer-based adjuvant composed of highly stable dispersion of microspherical particles of sodium polyacrylate in water (Figure 1A, Table 1). This polymeric technology has already been used in several vaccine models, including pet vaccines, with a promising safety and efficacy profile [19].

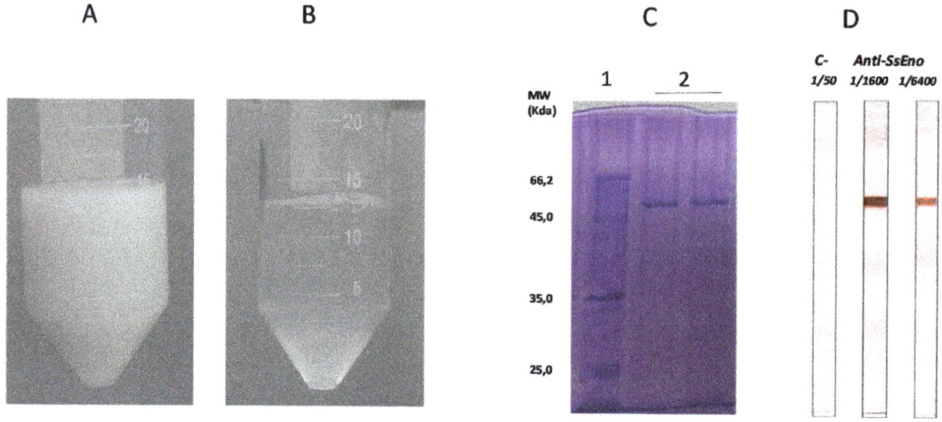


**Figure 1.** External aspect of (**A**) Montanide$^{TM}$ PetGel (PGA) and (**B**) PGA+rSsEno vaccine formulation. (**C**) SDS-PAGE showing the expression and purification of *rSsEno*. C1. Molecular pattern, C2. rSsEno band stained with Coomassie blue with the expected 47 kDa molecular mass (in duplicate). (**D**) Immunoblotting analysis of rSsEno with a pooled anti-*rSsEno* serum from Balb/c mice immunized with Freund's adjuvant/rSsEno (Either 1/1600 or 1/6400 dilutions of serum were used). A pooled serum from non-immunized mice (dilution 1/50) was used as negative control (C−) to identify non-specific binding.

**Table 1.** General properties of Montanide™ GEL adjuvants.

| | Description |
|---|---|
| Composition | Gel particles of sodium polyacrylate in water. |
| Particle size | 90% of the particles are smaller than 1.2 μm in diameter. |
| Stability | Highly stable at room temperature. |
| Mechanisms of action | Depot effect with slow release of antigens, due to polymer adsorption properties. Improves the recruitment and activation of the innate immune system cells and inducement of specific immune response. |
| Vaccine preparation | Montanide™ GEL adjuvants are ready-to-use adjuvants that can be combined with a wide range of antigens by gentle mixing. |
| Routes of administration | Parenteral and mucosal administration. |
| Uses | Montanide™ GEL adjuvants are recommended for a wide variety of livestock species and for pets and horses. They can be formulated with a wide range of antigens. |
| Safety | Montanide™ adjuvants and their components have been considered as safe by the Committee for Veterinary Medical Products (CVMP) for use in immunological products. They are included in Part I of the Annex of the European Council Regulation n° 37/2010/EU as substances needing no further MRL studies, in the Out of Scope list (EMACVMP-519714-2009), and included in already registered veterinary commercial products. |

Recently, our group evaluated a cell-wall protein extracted from *S. schenckii* (SsCWPs) in an experimental vaccine formulated with Aluminum hydroxide (AH) gel [20]. Immunized mice developed a specific immune response characterized by a balanced Th1/Th2/Th17 response and the production of protective antibody response. In a more recent study, the AH-based vaccine was compared with other experimental vaccine candidate containing SsCWPs and PGA. Both formulations induced a protective immune response in mice. However, AH stimulated the development of granulomas in the inoculation site while PGA-based vaccine exhibited a Th1 protective response and better local tolerance than AH-based vaccine in vaccinated mice [21]. In other recent study, we evaluated the immunogenic and protective effect of enolase, one of the main antigens that were found in SsCWPs, against *S. brasilensis*, the most virulent species of the genus *Sporothrix* [22]. The recombinant enolase of *S. schenckii* (SsEno) was formulated with PGA and after three subcutaneous administrations in mice, a significant specific immune response was observed. Furthermore, a reduction in mortality (over 90%) was observed after 45 days of an intravenous challenge with viable yeast, compared with non-vaccinated mice.

Until now, there are no studies on the effects of vaccination on fungal strains whose virulence changed due to exposure to environmental contaminants. In the current context of growing reports of fungal resistance to antifungal drugs and the emergence of new highly virulent strains, possibly acquired in the environment, these assessments can provide important information on the ability of new vaccines to prevent and combat emerging mycotic diseases.

In this study, we evaluated the protective capacity of an enolase-based vaccine formulated with PGA as an adjuvant, against *S. schenckii* with increased virulence by experimental exposure to toluene [7].

## 2. Materials and Methods

### 2.1. Animals

Male Balb/c mice (five to seven weeks old) were purchased from "Centro Multidisciplinar para Investigação Biológica na Área da Ciência de Animais de Laboratório" (CEMIB), Universidade de Campinas (UNICAMP), São Paulo, Brasil. Mice were housed in microisolator cages in a controlled ambient and receiving water and food ad libitum. The study was carried out in accordance with the Guide for the Care and Use of Laboratory Animals of the National Institutes of Health. The experiments were approved by the Ethics Committee for Animal Use in Research of Araraquara's School of Pharmaceutical Sciences from UNESP (Protocol CEUA/FCF/CAr: 19/2018).

### 2.2. Microorganisms and Preparations

The *S. schenckii sensu stricto* strain ATCC 16345 (here named as *S. schenckii*) used in this work was kindly provided by the Oswaldo Cruz Foundation (Rio de Janeiro, Brazil). The mycelial phase was maintained at room temperature in Mycosel (BD Biosciences) agar. A piece of a well-defined colony was grown in 100 mL of Sabouraud dextrose broth (SDB) (Difco, Detroit, MI, USA) for four days in a rotary shaker (130 rpm and 30 °C). The conidia were separated from the hyphae by filtration with sterile gauze using a Buchner funnel. Them conidia were counted and suspended in phosphate-buffered saline (PBS) at $1 \times 10^7$/mL.

### 2.3. *S. schenckii* Growth in Toluene

Fungal cultures were performed in 125-mL Erlenmeyer flasks containing 50 mL of Sabouraud dextrose broth (SDB) and sealed with Teflon Mininert valves (SUPELCO, 24 mm, (Merck KGaA Darmstadt, Germany) to prevent evaporation of the solvent. SDB was supplemented with toluene 0.1% ($v/v$). An aliquot of $1 \times 10^7$ conidia was inoculated and incubated during five days on a rotary shaker (30 °C and 130 rpm). Control cultures without toluene were included. Fungal viability was determined at fifth day by counting colony forming units (CFU) on Sabouraud dextrose agar (SDA) plates [7].

### 2.4. Expression and Purification of Recombinant *S. schenckii* Enolase (rSsEno)

The detailed procedures were previously described [22]. Briefly, the gene that encodes *S. schenckii* enolase with molecular mass 47 kDa and 438 amino acids (access code: ERS97971.1, GenBank database) was synthesized by Epoch Life Science Inc. (Missouri, TX, USA). The enolase gen was subcloned into the pET28a plasmid and optimized for production in *Escherichia coli* (pET28a::SsEno). *E. coli* DH5α was used for the propagation of pET28a::SsEno on lysogeny broth (LB) agar medium containing 30 μg/mL of kanamycin. For recombinant protein expression, *E. coli* BL21 cells cotransformed with pET28a::SsEno were grown at 37 °C in LB medium with kanamycin until they reached an OD600 in the range of 0.5–0.7. The expression of rSsEno was induced by 0.2 mmol/L of isopropyl β-D-1-thiogalactopyranoside (IPTG) at 30 °C for 4 h. The cells were centrifuged (20 min at 8000 rpm), and the pellet was resuspended in buffer A (NaPO4 20 mM, NaCl 500 mM and imidazole 20 mM, pH 7,4) containing 5 U of DNAse (Promega, Madison, WI, USA) and 30 μg/mL lysozyme (Merck KGaA Darmstadt, Germany) for 30 min on ice. The cell homogenate was sonicated, filtrated and then centrifuged at 19,000 rpm for 20 min at 4 °C. The supernatant containing rSsEno was filtered (0.45 μm nitrocellulose membrane, Millipore and initially purified by $Ni^{2+}$-affinity chromatography in buffer A. The rSsEno eluted in buffer B ($NaPO_4$ 20 mM, NaCl 500 mM, and imidazole 500 mM, pH 7.4) was subjected to size exclusion chromatography (SEC) with a Superdex 200 pg 16/60 column (GE Healthcare Life Sciences, Chicago, IL, USA) in Tris-HCl 25 mM, NaCl 100 mM and β-mercaptoethanol 2 mM at pH 7.5, and the eluted protein was concentrated using the Amicon® Ultra 15 mL 3k device (Millipore, Burlington, MA, USA) after being dialyzed for 24 h at 4 °C against phosphate buffer saline. The rSsEno concentration was measured by the Pierce BCA assay (Thermo Scientific, Waltham, MA, USA), and the efficacy of the expression and purification processes was assessed by 12% SDS-polyacrylamide gel electrophoresis (SDS-PAGE) and immunoblotting using anti-rSsEno serum.

### 2.5. Adjuvants and Vaccine Formulation

The vaccine formulation was prepared by mixing 100 μg of rSsEno with 5% PGA adjuvant kindly provided by Seppic (Paris, France) (Figure 1B). Other formulations composed by either PGA or rSsEno alone were used as control.

### 2.6. Immunization Schedule

Balb/c mice ($n$ = 5) received subcutaneous (s.c.) vaccination (on days 0 for priming and 14 for booster) in the back of the neck, with 100 μL of one of the following formulations: PGA+rSsEno, 100 μg rSsEno or PBS alone as a negative control. One week after the booster, mice were euthanized in $CO_2$ chamber and bled by heart puncture to obtain serum, which was aliquoted and stored at −20 °C until use.

### 2.7. Quantification of the rSsEno-Antibody Response by Enzyme-Linked Immunosorbent Assay (ELISA)

rSsEno IgG antibody titration was conducted as described previously [22]. Briefly, a 96-well ELISA plate (Merck KGaA Darmstadt, Germany) was coated with 5 μg rSsEno/mL in PBS and at 4 °C (overnight). The plate was washed with washing buffer (0.1% Tween 20) and then blocked 1 h at room temperature with 5% dried skim milk in washing buffer. Dilutions of the serum samples (1:500 in blocking buffer) were added to each well and incubated at room temperature for 2 h. After washing, peroxidase-conjugated anti-mouse IgG (1/500) (Merck KGaA Darmstadt, Germany) was added and incubated at 37 °C for 1 h. After exhaustive washing, tetramethylbenzidine was added to reveal the antigen-antibody reactions (30 min at room temperature). The reaction was stopped by the addition of 50 μL/well 1M $H_2SO_4$, and the absorbance was read with an ELISA reader (Multiskan Ascent, Labsystem, Vantaa, Finland) at 450 nm.

### 2.8. Th1-Th17 Phenotipagem

Spleens were aseptically removed and splenocytes were extracted. Viable splenocytes were adjusted to 1 × $10^7$ cells/mL in complete RPMI-1640 culture medium (Merck KGaA Darmstadt, Germany), which was supplemented with 2 mm L-glutamine, 100 U/mL penicillin, 100 μg/mL penicillin/streptomycin, and 10% fetal calf serum (RPMI complete). For study of Th1 and Th17 lymphocytes subpopulations, the following anti-mouse mAb were used: anti-CD16/CD32, anti-CD3-FITC, anti-CD4-APC, anti-IL-17-PE, anti-IFN-ɣ-Percp, and respective isotype controls (all purchased from BD Biosciences, (Franklin Lakes, NJ, USA). Splenocytes were assessed for the frequency of Th1(IFN-ɣ+), Th17 (IL-17+). Briefly, viable splenocytes were stained for the extracellular markers, then fixed and permeabilized using eBiosciences' intracellular fixation (Thermo Scientific, Waltham, MA, USA) and permeabilization buffer set, and then the intracellular IFN-ɣ and IL-17A were stained with a fluorescent respective marker. Intracellular cytokines were detected after in vitro stimulation with 10 μg/mL of rSsEno and Brefeldin A for intracellular retention of the induced cytokine. Events were acquired using a BD Accuri C6 flow cytometer (BD Biosciences) and analyzed with the flow cytometer's proprietary software.

### 2.9. IFN-ɣ, IL-4, and IL-17 Measurement in Supernatant of Splenocytes Culture

Splenocytes from immunized and non-immunized mice were cultured as previously described and stimulated with 10 μg/mL of rSsEno for 24 h. The levels of IFN-ɣ, IL-4, and IL-17 after rSsEno stimulation were measured in the supernatant of splenocytes culture by Cytometric Bead Array (CBA) (BD Biosciences) according to the manufacturer's instructions using a BD Accuri C6 flow cytometer (BD Biosciences).

### 2.10. Fungal Challenge and Infection Assessment

Either vaccinated or non-vaccinated mice were intraperitoneally inoculated with $10^6$ conidia of either wild type (WT) or toluene-adapted (Tadap) *S. schenckii* suspended in 100 μL of PBS or with an equal volume of PBS alone as control. To confirm the fungal cell count and viability of the inoculum, appropriately diluted samples of the conidia suspension were plated onto Mycosel agar plates and after seven days of incubation growing colonies were counted. At seventh day post-infection the mice were euthanized in $CO_2$ chamber. The liver and spleen of each animal were removed to measure the relative organ weight and assess the systemic fungal load. The relative weight of livers and spleens was calculated by the following formula:

$$\text{Related weight} = \text{organ weight (g)/body weight (kg)}.$$

To evaluate the fungal load, liver and spleen were macerated under sterile conditions and adequate dilutions of the macerate in PBS were cultured in duplicate, on Mycosel agar plates and the growing CFU were counted after three and six days. The final count was adjusted according to the dilution used.

### *2.11. Statistical Analysis*

Statistical analysis was performed in GraphPad Prism ver. 6.01 (San Diego, CA, USA). A one-way analysis of variance (ANOVA) with Tukey comparisons test was used. The confidence interval was set at 95% for all tests. The significance level and $p$-values were shown as * ($p < 0.05$); ** ($p < 0.01$); *** ($p < 0.001$); **** ($p < 0.0001$).

## 3. Results

### *3.1. Production and Purification of rSsEno*

Figure 1C,D show that the production and purification of rSsEno was effective. SDS-PAGE showed a unique band stained with Coomassie blue with the expected molecular mass of 47 kDa previously characterized as rSsEno [13]. In addition, a pooled anti-rSsEno serum obtained from mice immunized with Freund´s Adjuvants/rSsEno recognized specifically the rSsEno band in the immunoblotting while there were not detected non-specific binding in the control strip treated with a serum from non-immunized mice.

### *3.2. Post-Vaccination rSsEno-Antibody Response*

The rSsEno-specific IgG antibody reaction after the second immunization with or without the PGA are displayed in Figure 2 as optical density (OD) measured at 450 nm. Immunization with PGA+rSsEno markedly enhanced the IgG antibody response to rSsEno compared to the response induced by the non-adjuvanted vaccine, seven days after the second immunization ($p < 0.0001$).

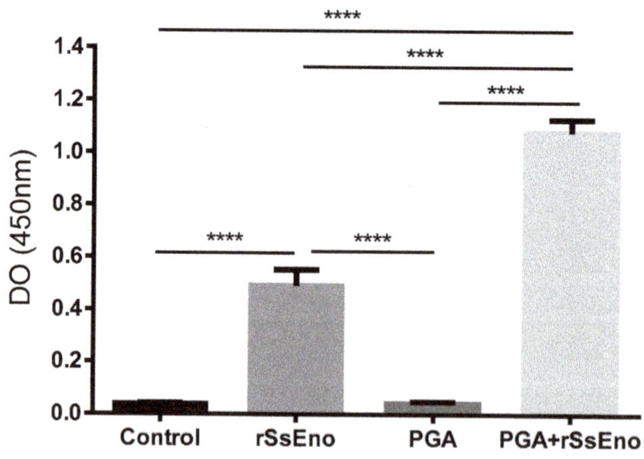


**Figure 2.** Immunization with PGA+rSsEno markedly enhanced the IgG antibody response to rSsEno. Balb/c mice were immunized (s.c.) twice with rSsEno, PGA, PGA+rSsEno, or PBS as negative control. Serum collected seven days after the second immunization was used to determine rSsEno-specific IgG antibody. **** ($p < 0.0001$).

*3.3. Th1 and Th17 Response*

The response Th1 and Th17 are determinant in the immune response against *S. schenckii*. Mice vaccinated with rSsEno+PGA induced a significant response of Th1 lymphocytes after in vitro stimulation with rSsEno compared with the other groups ($p < 0.0001$). However, the response of Th17 lymphocytes was not modified in any experimental group (Figure 3). A similar response was observed when the concentration of IFN-γ, IL-4, and IL-17A were measured by CBA. A high production of IFN-γ, belonging to the Th1 pattern was detected in the group vaccinated with rSsEno+PGA compared with non-adjuvanted *rSsEno* vaccination ($p < 0.001$) and with the control groups ($p < 0.0001$) (Figure 4).

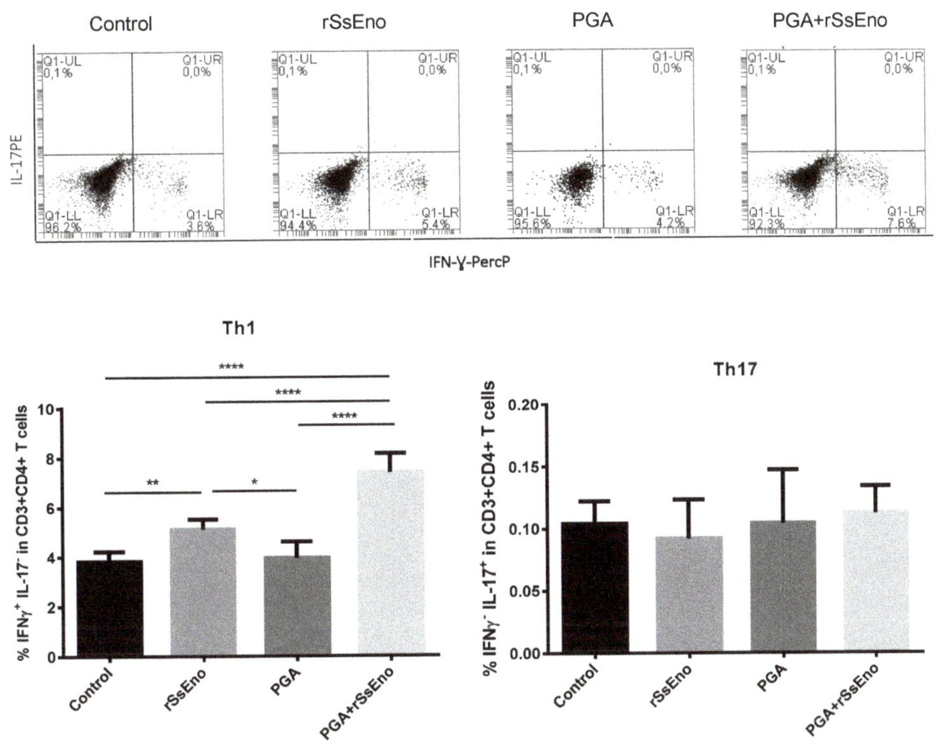


**Figure 3.** Response of Th1 (intracellular IFN-γ+) and Th17 (intracellular IL-17A+) cells induced by immunization with the PGA+rSsEno. Balb/c mice were immunized (s.c.) twice with rSsEno, PGA, PGA+rSsEno, or PBS as negative control. Splenocytes of each animal were collected seven days after the second immunization and purified cells were stimulated in vitro with rSsEno and Brefeldin A. The frequency of Th1 and Th17 cells was detected using a BD Accuri C6 flow cytometer. Upper figure shows representative dot plots of Th1 and Th17 frequency collected from each group. * ($p < 0.05$); ** ($p < 0.01$); *** ($p < 0.001$): **** ($p < 0.0001$).

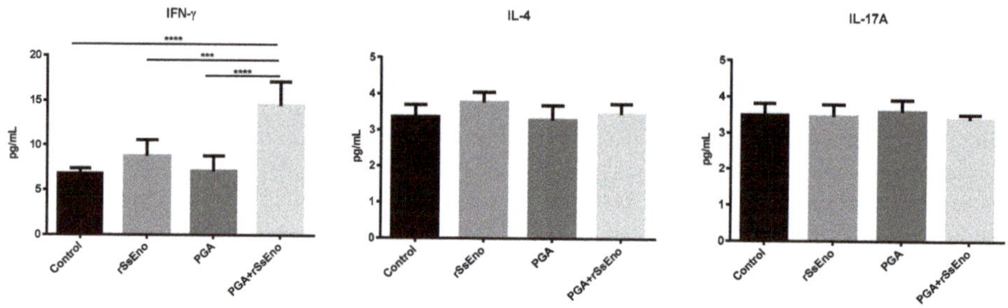


**Figure 4.** Cytokine profile in supernatant of splenocytes stimulated in vitro with rSsEno. Balb/c mice were immunized (s.c.) twice with rSsEno, PGA, PGA+rSsEno, or PBS as negative control. Splenocytes of each animal were collected seven days after the second immunization and purified cells were stimulated in vitro with rSsEno. Cytokine were quantified by cytometric bead array (CBA). * ($p < 0.05$); ** ($p < 0.01$); *** ($p < 0.001$): **** ($p < 0.0001$).

### 3.4. Fungal Challenge and Infection Assessment

The aim of this study was to investigate if the immune response induced by rSsEno adjuvanted with PGA was able to protect against a challenge with either WT or Tadap *S. schenckii*. We analyzed the spleen and the liver as representative organs to evaluate the fungal load. All the infected mice developed hepatomegaly that was observed by measuring the relative weight of the liver. However, splenomegaly was also observed in mice vaccinated and infected. Those animals vaccinated and infected with highly virulent toluene-adapted *S. schenckii* developed the greatest hepato- and splenomegaly compared with the control group ($p < 0.0001$) (Figure 5).

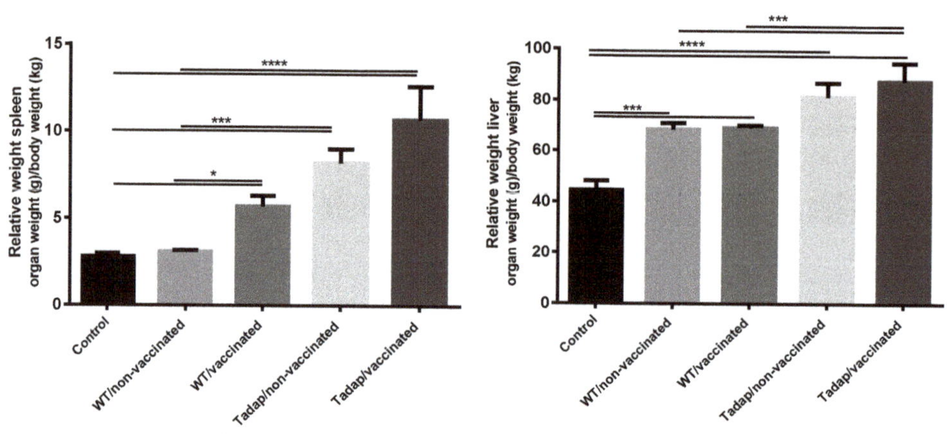


**Figure 5.** Relative (**A**) spleen and (**B**) liver weight. Balb/c mice were injected (s.c.) twice with PGA+rSsEno (vaccinated) or PBS (non-vaccinated). Seven days after the second dose of administration of PGA+rSsEno or PBS, mice were infected intraperitoneally with $1 \times 10^6$ conidia of either wild type (WT) or toluene adapted *S. schenckii* (Tadap). The spleen and the liver of each animal were collected seven days after the second immunization and the relative weight (organ weight/animal weight) was measured. * ($p < 0.05$); *** ($p < 0.001$); **** ($p < 0.0001$).

The protective effect induced by the adjuvanted vaccine after the fungal challenge is shown in Figure 6. The immune response induced by PGA+ rSsEno was able to reduce the fungal burden in spleen and liver of mice infected with either WT or Tadap *S. schenckii* in a similar way. However, owing to the higher virulence of Tadap *S. schenckii*, this finding suggests that PGA+rSsEno can be effective against fungus with different levels of virulence.

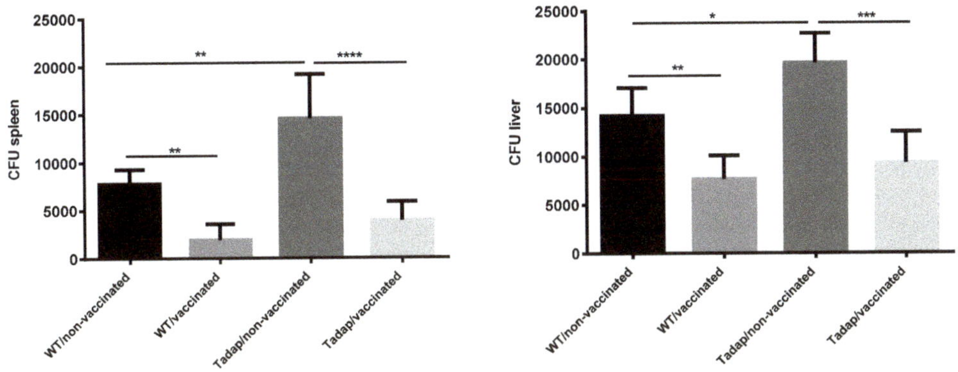


**Figure 6.** Vaccination with PGA+rSsEno was able to reduce the fungal burden in spleen and liver of mice infected with either wild type (WT) or toluene adapted *S. schenckii* (Tadap). Balb/c mice were injected (s.c.) twice with PGA+rSsEno (vaccinated) or PBS (non-vaccinated). One week after the boost, mice were i.p. challenged with *S. schenckii* and seven days after infection the protection was assessed by the number of CFUs recovered from the spleen and liver. * ($p < 0.05$); ** ($p < 0.01$); *** ($p < 0.001$); **** ($p < 0.0001$).

## 4. Discussion

Recent advances in the understanding of relevant immunological mechanisms against pathogenic fungi favor the development of prophylactic and therapeutic antifungal vaccines [23–25]. In contrast to classical antifungal medications, vaccines can be administered to large populations with low potential risks or side-effects [26]. Currently, there is a growing tendency to develop subunit vaccines, based in well-defined microbial components, in order to increase vaccine safety [27]. Unfortunately, pure antigens are poorly immunogenic, and they should be formulated with adjuvants to improve the vaccine efficacy [14,28].

Several non-toxic polymer adjuvants are being used for sustained delivery of protein subunit vaccines [29,30]. Polymeric adjuvants act through slow release of the antigen for the selective targeting to antigen presenting cells, promoting different signaling pathways including, activation of toll-like receptor(s) and inflammasome pathway or directly interacting with B cells. Polymer–antigen complex can be phagocytosed, and the antigen effectively presented to naive T cells via major histocompatibility complex (MHC) molecules [31].

Recently, we compared AH with the polymeric adjuvant PGA in a vaccine candidate against *S. schenckii* containing proteins extracted from the fungal cell wall. PGA induced a protective immune response against both *S. schenckii* and *S. brasiliensis* with lesser local toxicity than AH in vaccinated mice [21]. In other recent work, we developed recombinant enolase that was a key antigen in SsCWPs and it was evaluated in a vaccine formulated with PGA as an adjuvant. Again, protective immune response in vaccinated mice after a challenge with *S. brasiliensis* was observed with a survival rate above 90% after 45 days of intravenous fungal challenge [22].

Here, we evaluated the protective effect of PGA formulated with rSsEno against a highly virulent *S. schenckii*. The enhanced virulence was acquired after fungal exposure to 0.1% of toluene, which induced adaptive changes previously described, including enhanced melanosome formation and stronger antioxidant mechanism compared with the wild type strain [7].

Before the evaluation of the protective effect of vaccination, the immunogenicity of the vaccine after two subcutaneous doses of PGA+rSsEno was evaluated. In the aforementioned work [22], we evaluated the immunogenicity and efficacy of this vaccine candidate using three subcutaneous administrations and PGA 10%. However, here the immunogenicity of the vaccine was evaluated after two doses and PGA 5%, as we used in another vaccine candidate with SsCWPs [21]. As expected, the formulation of PGA+rSsEno induced elevated production of specific IgG antibodies and a strong Th1 response after in vitro stimulation of splenocytes with rSsEno. However, we did not observe a significant Th17 response analyzed by both intracellular and released IL-17A.

Several studies revealed that Th1 response plays a decisive role in the defense against *S. schenckii* infection [32–35]. In this sense, activation of Th1 lymphocytes is becoming an interesting immunomodulatory strategy against sporotrichosis. Flores-García et al. (2015) reported that treatment with recombinant murine IL-12 (rmIL-12) promotes Th1 immunity and clinical improvement in an experimental sporotrichosis gerbil model [36]. In other study, Batista-Duharte et al. (2016) evaluated the therapeutic effect of adjuvant Finlay cochleates 3 (AFCo3), a cochleate containing purified and non-toxic LPS derived from *Neisseria meningitidis B* as vehicle of Amphotericin B (AmB) to evaluate the combined effect of immunomodulation induced by AFCo3 and the antifungal effect in a murine model of *S. schenckii* infection. AFCo3 stimulated a strong Th1- and Th17 response associated with the antifungal effect of AmB, which significantly improved the fungal clearance [37]. Regarding vaccination, studies suggest that Th1 response is associated with anti-sporothrix vaccine protection [20–22,38,39] and the activation of dendritic cells by fungal wall proteins seems to be important in theTh1 bias [34]. However, the role of Th17 response to prevent sporotrichosis is controversial. Some studies reveal that anti-sporothrix vaccination can induce Th17- combined with Th1 response [20,38–40]. However, two studies using PGA as an adjuvant with either SsCWPs [21] or rSsEno [22], showed that Th1 response is sufficient to achieve protection against a challenge with *S. schenckii* or *S. brasiliensis*. In the previous comparative study, using SsCWPs as antigen with either AH or PGA as adjuvant revealed that the adjuvant is the main component that drives the quality of the immune response since AH induced a balanced Th1/Th2/Th17 response while PGA induced a Th1 response [21]. Interestingly, the reduction of fungal load in both groups was very similar despite having different overall Th cell pattern.

In this work, we observed that the anti- rSsEno Th1 response induced by PGA confers protection against Tadap *S. schenckii* after an intraperitoneal fungal challenge. Previously, we showed that infected mice with Tadap *S. schenckii* produced high levels of IFN-ɣ and, to a lesser extent, IL-17 [7]. This finding reinforces the criterion of the role of Th1-mediated response for protection against *S. schenckii* infection. Here, we observed significant production of IFN-ɣ after in vitro stimulation with rSsEno, compared to non-vaccinated mice, although the cytokine production was low in general. However, unlike infection where multiple lymphocyte clones are activated, vaccination stimulates specific clones, but they are sufficient in the booster phase to prevent infection. In this way, the role of vaccination is the expansion of specific lymphocytes to maintain a basal level of memory cell able to rapidly respond against microbial infection [41].

These results are important as part of the studies of the efficacy of the recombinant enolase-Montanide™ PetGel A vaccine, due to the environment can exert a direct influence on fungal virulence [4,5]. Ideally, a prophylactic vaccine should be able to prevent infections caused by microorganisms with different levels of virulence since imperfect vaccination can enhance the transmission of highly virulent pathogens [42]. In this sense, the choice of potent and safe adjuvants driving effective immunological mechanisms against the target microorganism is one of the most important factors for vaccine efficacy [28]. However, keeping in mind that besides the antigen and the adjuvant mode of action, vaccine efficacy and toxicity are strongly influenced by the genetic background of the host [43,44], additional studies in different mouse strains and non-rodent species will be performed.

## 5. Conclusions

In this work, we provide evidence that vaccination with a recombinant enolase-based vaccine formulated with PGA as an adjuvant promotes a protective immune response in mice, against a highly virulent *S. schenckii* by toluene exposure. To the best of our knowledge, this is the first experimental approach to assess the role of vaccination in providing protective immunity against pathogenic fungus with enhanced virulence by exposure to environmental contaminants. Further studies are necessary in order to evaluate the memory response, including other mouse strains and using a model of subcutaneous infection.

**Author Contributions:** Conceptualization, D.T.-M.; D.L.P.; A.B.-D.; I.Z.C. methodology, D.T.-M.; A.B.-D. D.L.P. M.L.L. formal analysis D.T.-M.; D.L.P.; A.B.-D.; investigation, D.T.-M.; A.B.-D. D.L.P., M.L.L.; resources, D.T.-M.; D.L.P.; A.B.-D.; data curation, A.B.-D.; writing—original draft preparation, D.T.M.; D.L.P.; A.B.-D.; writing—review and editing, D.T.-M.; D.L.P.; A.B.-D.; I.Z.; supervision, A.B.-D.; I.Z.C. project administration, I.Z.C.; A.B.-D.; funding acquisition, I.Z.C.

**Funding:** This work was supported by Fundacão de Amparo à Pesquisa do Estado de São Paulo (FAPESP, grants 2017/26774-3).

**Conflicts of Interest:** The authors declare no conflict of interest.

## References

1. De Beer, Z.W.; Duong, T.A.; Wingfield, M.J. The divorce of *Sporothrix* and *Ophiostoma*: Solution to a problematic relationship. *Stud. Mycol.* **2016**, *83*, 165–191. [CrossRef]
2. Gremião, I.D.; Miranda, L.H.; Reis, E.G.; Rodrigues, A.M.; Pereira, S.A. Zoonotic Epidemic of Sporotrichosis: Cat to Human Transmission. *PLoS Pathog.* **2017**, *13*, e1006077. [CrossRef] [PubMed]
3. Barros, M.B.; de Almeida Paes, R.; Schubach, A.O. *Sporothrix schenckii* and Sporotrichosis. *Clin. Microbiol. Rev.* **2011**, *24*, 633–654. [CrossRef] [PubMed]
4. Téllez, M.D.; Batista-Duharte, A.; Portuondo, D.; Quinello, C.; Bonne-Hernández, R.; Carlos, I.Z. *Sporothrix schenckii* complex biology: Environment and fungal pathogenicity. *Microbiology* **2014**, *160 Pt 11*, 2352–2365. [CrossRef]
5. Ramírez-Soto, M.C.; Aguilar-Ancori, E.G.; Tirado-Sánchez, A.; Bonifaz, A. Ecological Determinants of Sporotrichosis Etiological Agents. *J. Fungi* **2018**, *4*, 95. [CrossRef] [PubMed]
6. Batista-Duharte, A.; Téllez-Martínez, D.; Aparecida Jellmayer, J.; Leandro Portuondo Fuentes, D.; Campos Polesi, M.; Martins Baviera, A.; Zeppone Carlos, I. Repeated Exposition to Mercury (II) Chloride Enhances Susceptibility to *S. schenckii sensu stricto* Infection in Mice. *J. Fungi* **2018**, *4*, 64. [CrossRef]
7. Tellez-Martinez, D.; Batista-Duharte, A.; Silva, V.P.; Portuondo, D.F.; Ferreira, L.S.; Marisa Polesi, C.; Costa, C.; Carlos, I.Z. Adaptive stress response induced by toluene increases *Sporothrix schenckii* virulence and host immune response. *bioRxiv* **2019**. [CrossRef]
8. Kauffman, C.A.; Bustamante, B.; Chapman, S.W.; Pappas, P.G. Clinical practice guidelines for the management of sporotrichosis: 2007 update by the Infectious Diseases Society of America. *Clin. Infect. Dis.* **2007**, *45*, 1255–1265. [CrossRef]
9. De Almeida, J.R.F.; Kaihami, G.H.; Jannuzzi, G.P.; de Almeida, S.R. Therapeutic vaccine using a monoclonal antibody against a 70-kDa glycoprotein in mice infected with highly virulent *Sporothrix schenckii* and *Sporothrix brasiliensis*. *Med. Mycol.* **2015**, *53*, 42–50. [CrossRef]
10. Batista-Duharte, A.; Pereira, S.A.; Freitas, D.F.S.; Gutierrez-Galhardo, M.C.; Fuentes, D.P.; Carlos, I.Z. Therapeutic and Prophylactic Tools for Sporotrichosis: Current Strategies and Future Tendencies. In *Sporotrichosis: New Developments and Future Prospects*; Carlos, I.Z., Ed.; Springer: Cham, Switzerland, 2015; pp. 147–177.
11. Edwards, J.E. Fungal cell wall vaccines: An update. *J. Med. Microbiol.* **2012**, *6*, 895–903. [CrossRef] [PubMed]
12. Portuondo, D.L.; Ferreira, L.S.; Urbaczek, A.C.; Batista-Duharte, A.; Carlos, I.Z. Adjuvants and delivery systems for antifungal vaccines: Current state and future developments. *Med. Mycol.* **2015**, *53*, 69–89. [CrossRef]
13. Di Pasquale, A.; Preiss, S.; Tavares Da Silva, F.; Garçon, N. Vaccine Adjuvants: From 1920 to 2015 and beyond. *Vaccines* **2015**, *3*, 320–343. [CrossRef] [PubMed]
14. Pérez, O.; Batista-Duharte, A.; González, E.; Zayas, C.; Balboa, J.; Cuello, M.; Cabrera, O.; Lastre, M.; Schijns, V.E. Human prophylactic vaccine adjuvants and their determinant role in new vaccine formulations. *Braz. J. Med. Biol. Res.* **2012**, *45*, 681–692. [CrossRef] [PubMed]

15. Hendrick, M.J.; Goldschmidt, M.H.; Shofer, F.S.; Wang, Y.Y.; Somlyo, A.P. Postvaccinal sarcomas in the cat: Epidemiology and electron probe microanalytical identification of aluminum. *Cancer Res.* **1992**, *52*, 5391–5394. [PubMed]
16. Coffman, R.L.; Sher, A.; Seder, R.A. Vaccine adjuvants: Putting innate immunity to work. *Immunity* **2010**, *33*, 492–503. [CrossRef] [PubMed]
17. Batista-Duharte, A.; Téllez-Martínez, D.; Fuentes, D.L.P.; Carlos, I.Z. Molecular adjuvants that modulate regulatory T cell function in vaccination: A critical appraisal. *Pharmacol. Res.* **2018**, *129*, 237–250. [CrossRef] [PubMed]
18. Han, J.; Zhao, D.; Li, D.; Wang, X.; Jin, Z.; Zhao, K. Polymer-based nanomaterials and applications for vaccines and drugs. *Polymers* **2018**, *10*, 31. [CrossRef]
19. Deville, S.; Carneaux, E.; Bertrand, F.; Cauchard, S.; Cauchard, J.; Dupuis, L. Adjuvant formulation for companion animals vaccines. *Procedia Vaccinol.* **2011**, *4*, 104–112. [CrossRef]
20. Portuondo, D.L.; Batista-Duharte, A.; Ferreira, L.S.; Martínez, D.T.; Polesi, M.C.; Duarte, R.A.; de Paula, E.; Silva, A.C.; Marcos, C.M.; Almeida, A.M.; et al. A cell wall protein-based vaccine candidate induce protective immune response against *Sporothrix schenckii* infection. *Immunobiology* **2016**, *221*, 300–309. [CrossRef] [PubMed]
21. Portuondo, D.L.; Batista-Duharte, A.; Ferreira, L.S.; de Andrade, C.R.; Quinello, C.; Téllez-Martínez, D.; de Aguiar Loesch, M.L.; Carlos, I.Z. Comparative efficacy and toxicity of two vaccine candidates against *Sporothrix schenckii* using either Montanide™ Pet Gel A or aluminum hydroxide adjuvants in mice. *Vaccine* **2017**, *35*, 4430–4436. [CrossRef] [PubMed]
22. Portuondo, D.L.; Dores-Silva, P.R.; Ferreira, L.S.; Téllez-Martínez, D.; Marcos, C.M.; de Aguiar Loesch, M.L.; Guzman, Q.F.; Borges, J.C.; Batista-Duharte, A.; Carlos, I.Z. Immunization with recombinant enolase of Sporothrix spp (rSsEno) confers effective protection against sporotrichosis in mice. *bioRxiv* **2019**. [CrossRef]
23. Taborda, C.P.; Nosanchuk, J.D. Vaccines, Immunotherapy and New Antifungal Therapy against Fungi: Updates in the New Frontier. *Front. Microbiol.* **2017**, *8*, 1743. [CrossRef]
24. Nami, S.; Mohammadi, R.; Vakili, M.; Khezripour, K.; Mirzaei, H.; Morovati, H. Fungal vaccines, mechanism of actions and immunology: A comprehensive review. *Biomed. Pharmacother.* **2019**, *109*, 333–344. [CrossRef]
25. Spellberg, B. Vaccines for invasive fungal infections. *F1000 Med. Rep.* **2011**, *3*, 13. [CrossRef] [PubMed]
26. Di Pasquale, A.; Bonanni, P.; Garçon, N.; Stanberry, L.R.; El-Hodhod, M.; Tavares Da Silva, F. Vaccine safety evaluation: Practical aspects in assessing benefits and risks. *Vaccine* **2016**, *34*, 6672–6680. [CrossRef] [PubMed]
27. Moyer, T.J.; Zmolek, A.C.; Irvine, D.J. Beyond antigens and adjuvants: Formulating future vaccines. *J. Clin. Investig.* **2016**, *126*, 799–808. [CrossRef] [PubMed]
28. Batista-Duharte, A.; Martínez, D.T.; Carlos, I.Z. Efficacy and safety of immunological adjuvants. Where is the cut-off? *Biomed. Pharmacother.* **2018**, *105*, 616–624. [CrossRef]
29. Adams, J.R.; Haughney, S.L.; Mallapragada, S.K. Effective polymer adjuvants for sustained delivery of protein subunit vaccines. *Acta Biomater.* **2015**, *14*, 104–114. [CrossRef] [PubMed]
30. Shakya, A.K.; Nandakumar, K.S. Polymers as immunological adjuvants: An update on recent developments. *J. BioSci. Biotechnol.* **2012**, *1*, 99–210.
31. Shakya, A.K.; Nandakumar, K.S. Applications of polymeric adjuvants in studying autoimmune responses and vaccination againstinfectious diseases. *J. R. Soc. Interface* **2013**, *10*, 20120536. [CrossRef]
32. Uenotsuchi, T.; Takeuchi, S.; Matsuda, T.; Urabe, K.; Koga, T.; Uchi, H.; Nakahara, T.; Fukagawa, S.; Kawasaki, M.; Kajiwara, H.; et al. Differential induction of Th1-prone immunity by human dendritic cells activated with Sporothrix schenckii of cutaneous and visceral origins to determine their different virulence. *Int. Immunol.* **2006**, *18*, 1637–1646. [CrossRef]
33. Maia, D.C.; Sassá, M.F.; Placeres, M.C.; Carlos, I.Z. Influence of Th1/Th2 cytokines and nitric oxide in murine systemic infection induced by Sporothrix schenckii. *Mycopathologia* **2006**, *161*, 11–19. [CrossRef]
34. Quinello, C.; Ferreira, L.S.; Picolli, I.; Loesch, M.L.; Portuondo, D.L.; Batista-Duharte, A.; Carlos, I.Z. Sporothrix schenckii Cell Wall Proteins-Stimulated BMDCs Are Able to Induce a Th1-Prone Cytokine Profile In Vitro. *J. Fungi* **2018**, *4*, 106. [CrossRef] [PubMed]
35. Batista-Duharte, A.; Téllez-Martínez, D.; Roberto de Andrade, C.; Portuondo, D.L.; Jellmayer, J.A.; Polesi, M.C.; Carlos, I.Z. Sporothrix brasiliensis induces a more severe disease associated with sustained

Th17 and regulatory T cells responses than Sporothrix schenckii sensu stricto in mice. *Fungal Biol.* **2018**, *122*, 1163–1170. [CrossRef]

36. Flores-García, A.; Velarde-Félix, J.S.; Garibaldi-Becerra, V.; Rangel-Villalobos, H.; Torres-Bugarín, O.; Zepeda-Carrillo, E.A.; Ruíz-Bernés, S.; Ochoa-Ramírez, L.A. Recombinant murine IL-12 promotes a protective Th1/cellular response in Mongolian gerbils infected with Sporothrix schenckii. *J. Chemother.* **2015**, *27*, 87–93. [CrossRef]
37. Batista-Duharte, A.; Lastre, M.; Romeu, B.; Portuondo, D.L.; Téllez-Martínez, D.; Manente, F.A.; Pérez, O.; Carlos, I.Z. Antifungal and immunomodulatory activity of a novel cochleate for amphotericin B delivery against Sporothrix schenckii. *Int. Immunopharmacol.* **2016**, *40*, 277–287. [CrossRef] [PubMed]
38. De Almeida, J.R.F.; Jannuzzi, G.P.; Kaihami, G.H.; Breda, L.C.D.; Ferreira, K.S.; de Almeida, S.R. An immunoproteomic approach revealing peptides from Sporothrix brasiliensis that induce a cellular immune response in subcutaneous sporotrichosis. *Sci. Rep.* **2018**, *8*, 4192. [CrossRef]
39. Chen, F.; Jiang, R.; Wang, Y.; Zhu, M.; Zhang, X.; Dong, S.; Shi, H.; Wang, L. Recombinant Phage Elicits Protective Immune Response against Systemic S. globosa Infection in Mouse Model. *Sci. Rep.* **2017**, *7*, 42024. [CrossRef] [PubMed]
40. García-Lozano, A.; Toriello, C.; Antonio-Herrera, L.; Bonifaz, L.C. Sporothrix schenckii Immunization, but Not Infection, Induces Protective Th17 Responses Mediated by Circulating Memory CD4+ T Cells. *Front. Microbiol.* **2018**, *9*, 1275. [CrossRef]
41. Li Causi, E.; Parikh, S.C.; Chudley, L.; Layfield, D.M.; Ottensmeier, C.H.; Stevenson, F.K.; Di Genova, G. Vaccination expands antigen-specific CD4+ memory T cells and mobilizes bystander central-memory T cells. *PLoS ONE* **2015**, *10*, e0136717. [CrossRef] [PubMed]
42. Read, A.F.; Baigent, S.J.; Powers, C.; Kgosana, L.B.; Blackwell, L.; Smith, L.P.; Kennedy, D.A.; Walkden-Brown, S.W.; Nair, V.K. Imperfect Vaccination Can Enhance the Transmission of Highly Virulent Pathogens. *PLoS Biol.* **2015**, *13*, e1002198. [CrossRef] [PubMed]
43. Alba-Fierro, C.A.; Pérez-Torres, A.; Toriello, C.; Pulido-Camarillo, E.; López-Romero, E.; Romo-Lozano, Y.; Gutiérrez-Sánchez, G.; Ruiz-Baca, E. Immune Response Induced by an Immunodominant 60 kDa Glycoprotein of the Cell Wall of Sporothrix schenckii in Two Mice Strains with Experimental Sporotrichosis. *J. Immunol. Res.* **2016**, *2016*, 6525831. [CrossRef] [PubMed]
44. Batista-Duharte, A.; Portuondo, D.; Pérez, O.; Carlos, I.Z. Systemic immunotoxicity reactions induced by adjuvanted vaccines. *Int. Immunopharmacol.* **2014**, *20*, 170–180. [CrossRef] [PubMed]

*pharmaceutics*

MDPI

Article

# A Functional Wound Dressing as a Potential Treatment for Cutaneous Leishmaniasis

**Francisco Alexandrino-Junior [1], Kattya Gyselle de Holanda e Silva [2], Marjorie Caroline Liberato Cavalcanti Freire [3], Viviane de Oliveira Freitas Lione [2], Elisama Azevedo Cardoso [2], Henrique Rodrigues Marcelino [4], Julieta Genre [5], Anselmo Gomes de Oliveira [6] and Eryvaldo Sócrates Tabosa do Egito [1,5,7,*]**

1 Programa de Pós-Graduação em Nanotecnologia Farmacêutica (PPgNANOFARMA), Universidade Federal do Rio Grande do Norte (UFRN), Nata/RN 59012-570, Brazil; alexandrino_jr@yahoo.com.br
2 Faculdade de Farmácia, Universidade Federal do Rio de Janeiro (UFRJ), Rio de Janeiro/RJ 21941-902, Brazil; holanda.gyselle@gmail.com (K.G.d.H.eS.); vivianelione@gmail.com (V.d.O.F.L.); elisamaazevedo@gmail.com (E.A.C.)
3 Instituto de Física de São Carlos, Universidade de São Paulo (USP), São Carlos/SP 13563-120, Brazil; marjorie_freire_@hotmail.com
4 Departamento do Medicamento, Universidade Federal da Bahia (UFBA), Salvador/BA 40170-115, Brazil; henrique.rmarcelino@gmail.com
5 Faculdade de Farmácia, Universidade Federal do Rio Grande do Norte (UFRN), Nata/RN 59012-570, Brazil; jgenre@gmail.com
6 Departamento de Fármacos e Medicamentos, Universidade Estadual Paulista (UNESP), Araraquara/SP 14800-903, Brazil; ans_gomes@yahoo.com.br
7 Laboratório de Sistemas Dispersos (LaSiD), Departamento de Farmácia, Universidade Federal do Rio Grande do Norte (UFRN), Rua General Gustavo Cordeiro de Farias s/n, Petrópolis, Nata/RN 59012-570, Brazil
* Correspondence: socratesegito@gmail.com or socrates@ufrnet.br; Tel.: +55-84-99431-8816; Fax: +55-84-3342-9817

Received: 18 February 2019; Accepted: 21 March 2019; Published: 1 May 2019

**Abstract:** Cutaneous leishmaniasis (CL) is a parasitic disease characterized by progressive skin sores. Currently, treatments for CL are limited to parenteral administration of the drug, which presents severe adverse effects and low cure rates. Therefore, this study aimed to develop poly(vinyl-alcohol) (PVA) hydrogels containing Amphotericin B (AmB) intended for topical treatment of CL. Hydrogels were evaluated in vitro for their potential to eliminate promastigote forms of *Leishmania* spp., to prevent secondary infections, to maintain appropriate healing conditions, and to offer suitable biocompatibility. AmB was incorporated into the system in its non-crystalline state, allowing it to swell more and faster than the system without the drug. Furthermore, the AmB release profile showed a continuous and controlled behavior following Higuchi´s kinetic model. AmB-loaded-PVA-hydrogels (PVA–AmB) also showed efficient antifungal and leishmanicidal activity, no cytotoxic potential for VERO cells, microbial impermeability and water vapor permeability compatible with the healthy skin's physiological needs. Indeed, these results revealed the potential of PVA–AmB to prevent secondary infections and to maintain a favorable environment for the healing process. Hence, these results suggest that PVA–AmB could be a suitable and efficient new therapeutic approach for the topical treatment of CL.

**Keywords:** Amphotericin B; cutaneous leishmaniasis; hydrogel; wound dressing; controlled release

## 1. Introduction

Cutaneous leishmaniasis (CL) is a disease caused by a genus of trypanosomatid protozoa called *Leishmania*, transmitted to humans by the bite of infected female phlebotomine sandflies [1]. CL commonly appears first as a localized papule, which evolves into an ulcer upon loss of the epidermis. Afterwards, the impairment of the skin barrier in the lesion leads to the formation of long-life scars and severe skin disabilities. Moreover, CL, which represents the most common form of leishmaniasis, is currently considered a serious public health problem in 98 countries over all five continents [2]. The World Health Organization (WHO) has estimated that 0.7 to 1.3 million new cases of the disease occur worldwide annually [3].

While the development of an effective vaccine against *Leishmania* spp. is still under research [4,5], the use of pentavalent antimony organic compounds (SbV) or pentamidines are the recommended treatment for all leishmaniasis forms [6]. However, their efficiency for getting rid of the parasites is only around 60% [7]. In this context, Amphotericin B [8], an antifungal polyene agent approved by the FDA for clinical use, has been successfully applied when the abovementioned treatments failed, showing in some cases only 15% failure [9]. Nevertheless, the clinical use of AmB is limited due to the severe side effects, mainly nephrotoxicity, that the micelle system formulation containing this drug presents [8]. Besides, the use of AmB requires hospitalization for its intravenous administration, which leads to non-adhesion to the treatment by infected individuals in 75% of cases [7]. On the other hand, when less toxic AmB formulations are used, such as the liposomal ones [10], their cost is prohibitively expensive for people in developing countries, and these dosage forms still require intravenous administration.

Considering the present scenario, the development of new therapeutic approaches to improve CL treatment and to promote its world accessibility is mandatory. In this context, the topical treatment of CL lesions represents an attractive alternative to reduce the systemic toxicity associated with the use of the abovementioned dosage forms administered intravenously, promoting elimination of parasites, re-epithelization of the skin, preventing secondary infections and enabling outpatient treatment.

The currently available local treatments for CL include intralesional injection of SbV [11] or a combination of SbV with physical therapies, such as cryotherapy [12] or thermotherapy [13]. Additionally, the topical administration of paromomycin-methylbenzethonium chloride (PR–MBCL) ointments is an option. However, although this last treatment presents fewer adverse effects, less pain, and easier administration, its therapeutic activity depends on the presence of MBCL in the formulation [14], a cationic surfactant which usually leads to inflammatory reactions. Moreover, Kim et al. demonstrated that PR regimen was less effective than any SbV regimen to achieve a clinical cure, with a described efficacy varying from 17% to 67% for *Leishmania major* infection [14].

Undeniably, the development of topical formulations containing AmB seems to be the ideal approach for CL treatment. However, the commercially available lipid formulations for this drug, when topically applied, are ineffective at curing CL in an animal model [15]. This observation supports the idea that the lack of efficiency of this type of treatment could be a deficiency of the drug delivery rather than a lack of drug efficacy. Therefore, this highlights the need to develop new therapeutic systems loaded with AmB for the topical treatment of CL.

In this context, the use of hydrogels seems to be a promising strategy, since it combines numerous advantages to wound management, e.g., enhancing the healing process [16,17], reducing pain [18], enabling exchange of gases (e.g., $O_2$ and $H_2O$) [19], possibility to tailor the mucoadhesion [20], act as a barrier to external threats like microbes [19]. Moreover, it can, simultaneously, act as a carrier for therapeutic agents [21–24].

Hydrogels consist of polymeric networks that absorb large amounts of water while remaining insoluble in aqueous solutions due to chemical or physical cross-linking of their individual polymer chains. Since the first report in 1960, by Wichterle and Lím [25], synthetic hydrogels have been widely applied to biomedical use, and special attention should be given to the ones manufactured with poly(vinyl alcohol) (PVA). Due to its excellent properties (e.g., high biocompatibility, hydrophilicity, transparency, etc.) this polymer has a historical use in biomedical applications [26], e.g., cell culturing [27], artificial cartilage [28], long-term implants [29,30], scaffold for tissue engineering and tissue mimicking [31], wound dressing material [32,33] and soft contact lenses [34]. Lastly, PVA has been used as carriers for therapeutic agents such as drugs [35–39] and as a surface modifier, for liposomes [40,41]. The use of PVA in the pharmaceutical field, as a material to manufacture drug delivery systems, is justified because of its biodegradability [42] and low toxicity [43]. However, due to its high water solubility, PVA needs to be subjected to a cross-linking process in order to manufacture the hydrogel system. To this end, chemical cross-linking is one of the approaches used to enhance its mechanical, chemical and thermal properties [44].

Recently, our group demonstrated the feasibility of PVA hydrogels as a carrier to control AmB release [35]. Thus, the proposal of the present work was to produce an AmB-loaded poly(vinyl alcohol) hydrogel (PVA–AmB) intended to be used as a wound dressing system for CL treatment. With this aim, a $2^3$ full factorial design was applied to evaluate the role of the pH, the degree of cross-linking and the presence of AmB in the system´s constitution. Then, the hydrogel microstructure was characterized by evaluating the AmB aggregation state into the system, surface morphology, swelling degree, drug release kinetics, and water and microbial permeability. The final goal of this proposal was to assess the leishmanicidal, antifungal, and cytotoxic activity of the AmB-loaded system.

## 2. Materials and Methods

### 2.1. Materials

Glutaraldehyde (GA, 25%*v*/*v* aqueous solution), dimethyl sulfoxide (DMSO) and monobasic anhydrous potassium phosphate were purchased from Quimiobras Indústria Química (Masssaranduba, SC, Brazil). Sodium chloride, potassium chloride, and methanol came from ISOFAR Indústria e Comércio de Produtos Químicos (Duque de Caxias, RJ, Brazil). Amphotericin B came from Indofine Chemical Company (Hillsborough, NJ, USA). Brain heart infusion medium and Sabouraud Dextrose Agar were purchased from HiMedia Laboratories, LLC (Mumbai, India). Streptomycin and penicillin were purchased from LGC Biotecnologia (Cotia, SP, Brazil). Fetal bovine serum (FBS) was purchased from Gibco™ (Gaithersburg, MD, USA). Dibasic sodium phosphate, Schneiders modified medium, Dulbecco´s Modified Eagle´s Medium–High glucose (DMEM–HG), and anhydrous poly(vinyl alcohol) (PVA, 98% hydrolyzed, Mw 13.000–23.000 g/mol) were purchased from Sigma-Aldrich (St. Louis, MO, USA).

### 2.2. Factorial Design for Development of AmB-Loaded Hydrogels

A $2^3$ full factorial design was used to evaluate the influence of three main factors on the system's characteristics. In this case, the pH (5 and 2), the cross-linker concentration (Glutaraldehyde at 132.5 $\mu M \cdot L^{-1}$ and at 26.5 $\mu M \cdot L^{-1}$), and the presence of the AmB in the system (0 and 10 μM) were tested.

As previously reported, PVA hydrogels were prepared using the casting technique with few modifications [45]. Briefly, PVA powder was accurately weighed, introduced into an airtight container with water, and completely dissolved at 90 °C for 15 min to form a 10%(*w*/*v*) solution. This solution was cooled slowly at room temperature, under magnetic stirring, and the pH was adjusted to 2 with HCl 1 M. Afterward, AmB was added to the PVA solution in a 1:20 ratio (AmB:PVA *w*/*w*), followed by ultrasound bath for 10 min. The cross-linking reaction was carried out for 16 h in a polyethylene dish (diameter of 5.5 cm). Before all analyses, except the X-ray diffraction and the scanning electron microscopy, the hydrogels were immersed twice in distilled water to remove $H^+$ excess.

### 2.3. X-ray Diffraction Study

The X-ray diffraction (XRD) spectrum of the hydrogels with (PVA–AmB) and without AmB (PVA–H), respectively, were carried out in the range from 2° to 80° (2θ) at the speed of 0.05°/s using a diffractometer (Miniflex desktop diffractometer, Rigaku Corporation, Tokyo, Japan). The XRD system was equipped with a CuKα radiation source (λ =1.541 Å, 40 kV and 40 mA).

### 2.4. Hydrogel Thickness and Morphology

In order to investigate the thickness homogeneity in the PVA–AmB, the systems were virtually divided into two regions, edge and inner. The thickness was determined using a manual ABS digimatic indicator (model 543250B /ID-C112B, Mitutoyo Corporation, Kanagawa, Japan). The final results were expressed as the mean of 20 measurements on the entire film surface (10 at random edge and inner regions, respectively).

The scanning electron micrographs of PVA–H and PVA–AmB were performed using a Scanning Electron Microscope (JSM-5610LV Scanning Electron Microscope, JEOL Ltd, Tokyo, Japan). Prior to evaluation, hydrogels were placed over carbon conductive tabs and sputter-coated with gold plasma. The morphology of the system and the presence of pores were evaluated using an accelerating voltage of 20 kV.

### 2.5. Drug Loading Efficiency (%)

The amount of AmB presented in the system was measured from hydrogel fragments of 1 cm$^2$. To this end, these pieces were immersed in 3 mL of DMSO and taken to the ultrasound bath for 15 min. The resulting solution was again diluted in DMSO and the AmB content was measured by spectrophotometry (Libra S32 UV/Vis Spectrophotometer, Biochrom Ltd, Cambridge, UK) at $\lambda = 416$ nm. Drug loading efficiency was calculated by the following equation [46]:

$$Drug\ Loading\ (\%) = \frac{Experimental\ Drug\ Content}{Theoretical\ Drug\ Content} \times 100 \tag{1}$$

### 2.6. Determination of Swelling Behavior

A gravimetric approach was used in order to measure the swelling behavior of the system. Therefore, PVA–H and PVA–AmB hydrogels were cut into discs of 1 cm in diameter and immersed into phosphate buffer saline (PBS, pH 7.3 at 25 °C), which was prepared according to the Brazilian Pharmacopoeia [47], for a pre-determined time. Moreover, the samples were taken out of the swelling medium and weighed again after removing PBS excess. The swelling degree ($Q$) of the hydrogel at each time was calculated as follows [48]:

$$Q = \frac{M_t - M_0}{M_0} \times 100 \tag{2}$$

where $M_t$ and $M_0$ are the weight of the system imbibed with the swelling medium at time $t$ and dry sample, respectively.

### 2.7. In Vitro Drug Release

Since the AmB has a very low water solubility (<1 mg·L$^{-1}$ at pH 6–7) [49], the addition of a solubility enhancer is a usual approach on drug delivery. Therefore, in order to balance the duality of performing changes in the release medium and, at the same time, still produce reliable results, the experiments were performed using the release medium previously reported in the literature [50].

The total immersion method was used to evaluate the in vitro AmB release from the system. To this end, hydrogels were cut into fragments of 1 cm$^2$ (equivalent to 216 ± 20 µg of AmB) and immersed in 150 mL of release medium (PBS:Methanol (80:20 *v*/*v*)) pH 7.3 at 37 °C, and stirred at 100 rpm. At specified time intervals, aliquots of 3 mL were withdrawn from the solution and the amount of AmB was measured by spectrophotometry (Libra S32 UV/Vis Spectrophotometer, Biochrom Ltd, Cambridge, UK) at $\lambda = 416$ nm with a previously validated method. The amount of drug released was calculated and represented as the accumulative percentage of the drug released versus time. To maintain sink conditions, each aliquot was replaced by the same volume of fresh medium.

### 2.8. Mathematical Analysis of the In Vitro Release Kinetics

Different mathematical models were evaluated in order to describe the mechanism of the AmB release that best suited the hydrogels, according to the experimental data obtained from the in vitro release assays (Table 1). Calculations were performed with the support of Add-in DDsolver for Microsoft® Excel [51]. The choice of the model took into account the statistical significance of the fitting and the thermodynamic considerations of the model.

**Table 1.** Different mathematical models used for fitting experimental data and their equations with the values of the statistical parameters.

| [a] Model | Equation | R2_adj | RMSE |
|---|---|---|---|
| [b] Zero-order | $F = k_0 \times t$ | 0.87 | 9.8 |
| [c] First-order | $F = 100 \,.\, (1 - e^{-k_1 \times t})$ | 0.97 | 4.7 |
| [d] Quadratic | $F = 100 \times (k_1 \times t + k_2 \times t^2)$ | 0.98 | 4.1 |
| [e] Higuchi | $F = k_H \times t^{0.5}$ | 0.93 | 7.2 |
| [*e] Higuchi with Tlag | $F = k_H \times (t - t_{lag})^{0.5}$ | 0.99 | 3.1 |
| [f] Baker–Lonsdale | $\frac{3}{2}\left[1 - \left(1 - \frac{F}{100}\right)^{2/3}\right] - \frac{F}{100} = K_{BL} \times t$ | 0.87 | 9.9 |
| [*f] Baker–Lonsdale with Tlag | $\frac{3}{2}\left[1 - \left(1 - \frac{F}{100}\right)^{2/3}\right] - \frac{F}{100} = K_{BL} \times (t - T_{lag})$ | 0.99 | 2.3 |
| [g] Korsmeyer–Peppas | $F = k_{KP} \times t^n$ | 0.5 | 16.5 |
| [*g] Korsmeyer–Peppas with Tlag | $F = k_{KP} \times (t - T_{lag})^n$ | 0.48 | 17.7 |
| [h] Hopfenberg | $F = 100 \times [1 - (1 - k_{HB} \times t)^n]$ | 0.94 | 6.47 |
| [*h] Hopfenberg with Tlag | $F = 100 \times \{1 - [1 - k_{HB} \times (t - T_{lag})^n]\}$ | 0.94 | 6.85 |

[a] In all mathematic models $F$ represents the fraction (%) of the drug released over time $t$; [b] $k_0$ = zero-order release rate constant; [c] $k_1$ = first-order release rate constant; [d] $k_1$ = release rate constant for the Quadratic model denoting the dependence of the drug release on the time; $k_2$ = release rate constant for the Quadratic model denoting the dependence of the drug release on the quadratic time; [e] $k_H$ = Higuchi release constant; [f] $K_{BL}$ = combined constant on the Baker–Lonsdale model; [g] $k_{KP}$ = release constant incorporating structural and geometric characteristics of the drug-dosage form; $n$ = diffusional exponent indicating the drug-release mechanism; [h] $k_{HB}$ = combined constant on Hopfenberg model; $n$ = 1, 2, and 3 for a slab, cylinder, and sphere, respectively; * $T_{lag}$ = lag time prior to drug release.

### 2.9. Water Vapor Transmission

The water vapor transmission (WVT) of hydrogels was determined using a modified ASTM E96/E96M water method [52]. Briefly, a test tube containing distilled water was sealed with the hydrogel. To avoid water transport through the edge, the test tubes were thoroughly sealed with scotch tape. Then, the assembly was placed into a chamber at 25 ± 1 °C with a constant relative humidity of 33% ± 1%. Finally, the change in weight of the assembly was measured and the rate of WVT was calculated using the following equation [53]:

$$WVT = \frac{W}{A \times \Delta p} \quad (3)$$

where $W$ is the amount of water vapor permeating through the hydrogel (g·day$^{-1}$), $A$ is the area of exposed hydrogel (cm$^{-2}$), and $\Delta p$ is the vapor pressure difference (mmHg).

### 2.10. Microbial Permeability Assay

In order to investigate the ability of the hydrogel to prevent microbial penetration, microbiological tests were conducted as previously mentioned [19], with modifications. Briefly, test tubes containing 3 mL of sterile brain heart infusion medium were sealed with sterile hydrogel at aseptic conditions. Positive and negative controls were an open and cap-closed test tube, respectively. The assembly was placed in an open environment and the progress of microbial permeation was observed for 7 days. The cloudiness of the medium in any test tube was recorded as microbial contamination.

### 2.11. Leishmanicidal Activity Assay

In order to estimate the leishmanicidal activity of the PVA–AmB, in vitro assays were performed as previously reported [54] with modifications. Briefly, PVA–H and PVA–AmB hydrogels were cut into fragments of 1 $cm^2$ and immersed into Schneiders modified medium. Afterward, late log-phase promastigotes of *Leishmania amazonensis* (IOC/L0575(IFLA/BR/1967/PH8)) and *Leishmania braziliensis* (IOC/L0566(MHOM/BR/1975/M2903) were added to the culture medium in the concentration of $1 \times 10^5$ parasites·$mL^{-1}$ and cultured in 24-well plates at 28 °C under 5% $CO_2$ for 48 h. At specific time intervals, the promastigote viability was microscopically determined at 400× magnification, through flagellar motility, and counted in a Neubauer chamber. The experiments were carried out under similar conditions using non-treated cells and free AmB (equivalent to a final concentration of 50 μg/mL) as negative and positive controls, respectively.

### 2.12. Antifungal Activity

The disc diffusion approach was used to evaluate the pharmacological efficacy of the PVA–AmB hydrogel against *Candida albicans* (ATCC®: 10231), as previously reported [55] with modifications. Briefly, a suspension of this fungal strain, equivalent to 0.5 on the McFarland scale, was homogeneously distributed on a Petri dish with Sabouraud Dextrose Agar using a sterile swab. Afterward, PVA–H and PVA–AmB hydrogels were cut aseptically in fragments of 1 $cm^2$ and neutralized with PBS. Then, hydrogels were put on the surface of the culture medium and the zones of growth inhibition were measured after 24 and 48 h of incubation at 28 °C. Control experiments were carried out under similar conditions using 10 μL of an AmB dispersion in PBS (2 mg/mL) as the standard drug with antifungal activity. Microorganism sensitivity to PVA–H and PVA–AmB hydrogels was determined by measuring the size of the inhibitory zones on the agar surface around them.

### 2.13. Cytotoxicity Assay

The nephrotoxicity of the hydrogels was estimated performing an in vitro assay. Firstly, African Green Monkey Kidney (VERO, ATCC® CCL-81) cells were seeded at $10^4$ cells per well and cultured in Dulbecco´s Modified Eagle´s Medium-High glucose (DMEM–HG) supplemented with 10% FBS, streptomycin (100 mg/mL), and penicillin (100 UI/mL). When the cultures reached confluence, transwell filters (0.8 μm; BD) containing PVA–H or PVA–AmB hydrogel fragments of 0.5 $cm^2$ (AmB ≈ 113 μg) were applied. Then, the plates were incubated at 37 °C under 5% $CO_2$ for 24 h. After treatment, the cytotoxicity was evaluated by the MTT cell proliferation assay [56]. The viability of non-treated control cells was defined as 100%.

In order to better understand the main effect of AmB toxicity to VERO cell line, a kill curve was performed. To this end, a stock solution of AmB in DMSO was diluted into the culture medium to obtain a concentration of AmB in the range from 5 to 50 μg/mL. The cells were cultivated according to the above-mentioned methodology.

### 2.14. Statistical Analysis

Statistical analysis was performed by the GraphPad Prism 5.03 software (GraphPad Software Inc, San Diego, CA, USA) using the Shapiro–Wilk test to evaluate the normality of the data distribution, followed by the Bartlett test to assess the homogeneity of variance. Afterward, unpaired *t* tests were performed to determine the difference between the means. The results were presented as the mean of three individual experiments ± standard error of the mean, with $p$-value < 0.05 considered significant.

## 3. Results and Discussion

### *3.1. Factorial Design*

The influence of the pH over the system's characteristics was considered significant once only hydrogels produced at pH 2 were able to maintain their structure during the dissolution test in the aqueous medium, while those produced at pH 5 were completely dissolved. Furthermore, the cross-linker concentration also influenced the system's structure. A synergism during the cross-linking reaction was observed with a higher level of both AmB (at 10 μM) and GA (at 132.5 $\mu M{\cdot}L^{-1}$), showing excessive cross-linking bonds, and leading to significant deformations of the hydrogel. This was an unexpected finding and it was ascribed to an increase in the number of hydrophilic interactions among the polymer chains, the cross-linker, and the AmB (Figure 1A), improving the physical and chemical cross-linking occurrence [57,58].

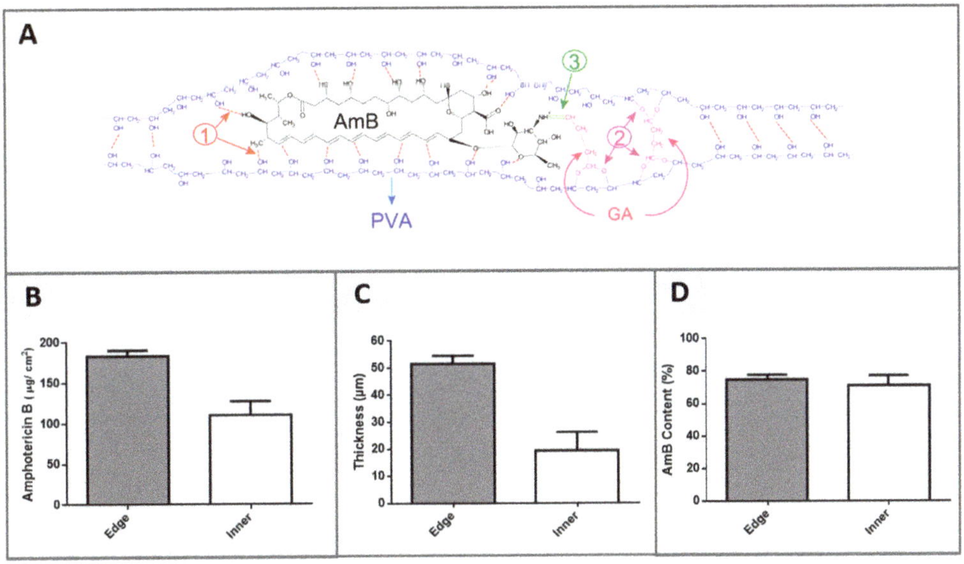


**Figure 1.** (**A**) Schematic representation of the hydrophilic interactions among the polymer chains, the cross-linker, and the Amphotericin B (AmB). (1) Hydrogen bound, (2) acetal formation and (3) imine formation improving the physical and chemical cross-linking [59]. (**B**) AmB content per unit area (1 $cm^2$), (**C**) thickness variation according to the hydrogel region (**D**) AmB values normalized according to the weight of fragments from each respective region.

### *3.2. Drug Loading Efficiency (%)*

The efficiency of AmB loading into the system was evaluated, as well as a possible difference in the drug distribution between the edge and the inner regions of the hydrogel. In fact, the quantitative analysis of the distribution of the drug into hydrogel showed a significant higher concentration of AmB at the film edges ($p$-value < 0.05), when compared to its inner region (Figure 1B). This result was ascribed to the capillarity phenomenon, which occurred during the cross-linking reaction at the border of the Petri dish used to generate the system, hence increasing the AmB content in this region.

This hypothesis was corroborated by the thickness comparison and PVA–AmB ratio of both hydrogel regions, inner and edge (Figure 1C). Indeed, the edge region was approximately 2.6 times thicker than the inner region ($p$-value < 0.05), but the PVA–AmB ratio did not change. In fact, the AmB content was normalized according to the weight of each respective fragment and showed no statistical difference between these regions ($p$-value > 0.05, Figure 1D).

Taking into account that the WHO [6] recommends the collection of samples from the swollen edge of CL ulcers for diagnostics due to the highest number of parasites found in this part of the lesion, the heterogeneity observed in the AmB distribution inside the hydrogel can be exploited as a therapeutic strategy. This could assure a better management of the administrated dose, targeting the majority of the active compound to sites where more parasites are found, leading consequently to a more efficient therapy.

### 3.3. Swelling Behavior

The swelling behavior of the hydrogel is a dynamic process, which depends on structural factors such as geometry, chemical composition of the polymeric network, the presence of pores and polymer–solvent interactions [60–63]. As AmB release from the hydrogel requires the previous solvation of the polymeric chains, swelling kinetic experiments were performed to better understand the mechanisms involved in this process.

Although there was no statistical difference in the swelling behavior of the hydrogels, Figure 2A shows that the PVA–AmB hydrogel had the ability to absorb four times its weight in water. This property could be of great applicability in the treatment of the leishmaniotic ulcers, as it may assist in the removal of secretions, which are presented in up to 86.9% of the lesions [64], and could be extremely beneficial for the healing process.

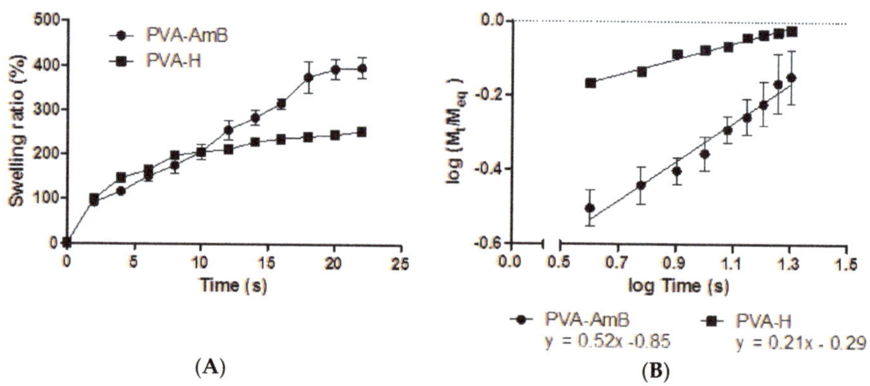


**Figure 2.** (**A**) Swelling degree of poly(vinyl-alcohol) (PVA)–H and PVA–AmB. (**B**) Log-log plot from which were calculated the diffusional exponent ($n$) and the diffusion constant ($k$) of PVA–H and PVA–AmB in phosphate saline buffer.

In addition, Ritger and Peppas [65] showed that from the swelling kinetic assay data it is possible to determine the mechanisms involved in the release process mainly by the use of the following equation:

$$\frac{M_t}{M_{eq}} = kt^n \tag{4}$$

where $n$, $k$, $M_t$ and $M_{eq}$ represents the diffusional exponent, the diffusion constant, and the mass of the hydrogel at the time $t$ and at the equilibrium, respectively. The values of these constants may be determined by Plotting $\log(\frac{M_t}{M_{eq}}) \times \log(t)$ (Figure 2B), where the angular and linear coefficients correspond to $n$ and $k$, respectively [63].

Although the presence of AmB did not change the swelling degree ($p > 0.05$), its presence significantly altered the sorption mechanism. The n values for PVA–H and PVA–AmB hydrogels were 0.2 and 0.52, respectively, which indicate a change in the swelling mechanism, from the Less Fickian to the Limited Relaxation of the polymeric chains one (Anomalous Process) [65]. This hindrance solvent permeation observed for the PVA–H hydrogel was attributed to the strong inter- and intra-chain interactions between PVA and GA in the absence of AmB. This fact can be evidenced by the analysis of the diffusion coefficient values of the solvent into the hydrogels, which were $11.10^{-2}$ and $17.10^{-2}$ $cm^2 \cdot s^{-1}$ for PVA–H and PVA–AmB, respectively. Then, the presence of AmB in the hydrogel increased approximately 1.5 times the speed of solvent permeation into the system microstructure.

### 3.4. *In Vitro Drug Release*

The evaluation of the AmB kinetic release profile from the hydrogel showed that, despite the rapid swelling, the AmB release occurred slowly and gradually. After 97 h, approximately 74% of the total drug content was released from the system. Afterward, the data were fitted using different mathematical models previously described in the literature [51]. The statistical parameters used to compare these different models were the Adjusted Coefficient of Determination ($R^2$_adj), which allowed comparing the fitting of the theoretical models to the experimental data, and the Root Mean Square Error (RMSE), which evaluated the difference between the experimental obtained data and the fitted data provided by the used model. Consequently, the model that best fit the experimental data had to exhibit the largest $R^2$_adj and the lower RMSE (Table 1). Although from the statistical point of view the Baker–Lonsdale with Tlag was the model that best described the experimental data, from the thermodynamic point of view this model was inappropriate since this approach, adapted from the Higuchi model, is applied to systems having spherical geometry such as microspheres or microcapsules. The literature reports that when the mechanistic models are under investigation, its selection should be based not only on the parameters that fit best, but also on the mechanistic model probability [51]. Thus, the model of Higuchi with Tlag was the chosen model (Figure 3A). This model (Table 1) depicts the release mechanism as a diffusion process based on Fick's law and dependent on the square root of time.

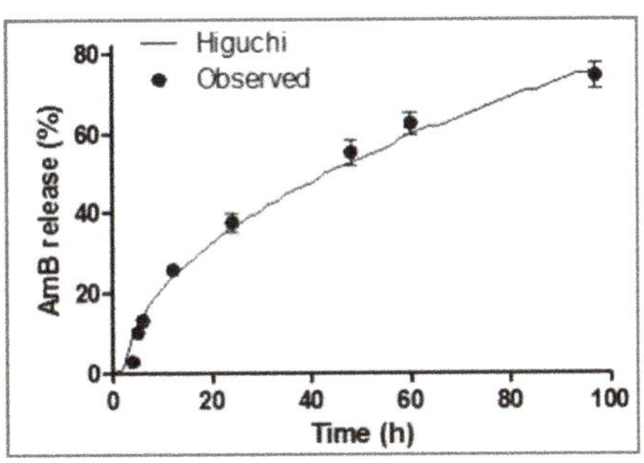


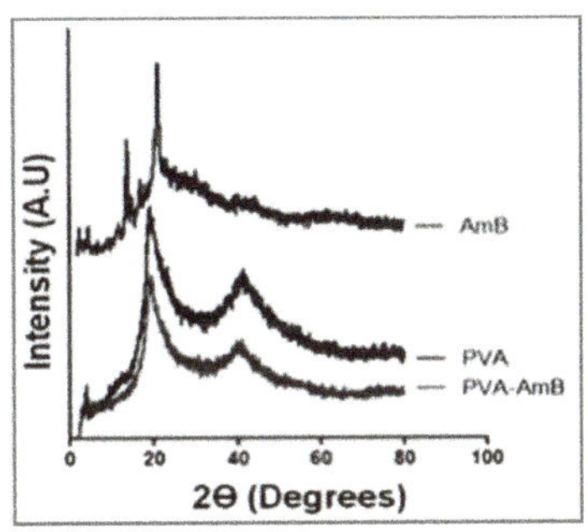


**Figure 3.** (**A**) Kinetic release of AmB from PVA–AmB hydrogels. Key: (•) experimental data (–) fitted data according to the Higuchi model. (**B**) XRD patterns of the AmB, the PVA–H, and the PVA–AmB hydrogels.

Another result that corroborates the hypothesis of the AmB release from the hydrogel being governed primarily by diffusion is the system microstructure organization inferred by the XRD analysis (Figure 3B). The diffractogram showed an absence of crystalline peaks ascribed to the AmB, indicating that the AmB is in an amorphous state or molecularly dispersed in the hydrogel. Additionally, the crystallinity index (CI) of the polymeric network calculated through the equation below, was 16.8% and 17.5% for PVA–H and PVA–AmB hydrogels, respectively, and showed no significant changes.

$$CI_{(\%)} = \left(1 - \frac{A_a}{A_t}\right) \times 100 \quad (5)$$

where $A_a$ and $A_t$ are the amorphous phases and the total area, respectively.

Furthermore, the release by erosion mechanisms, as previously reported for PVA hydrogels [66], had not fitted the data, which is justified by the absence of groups subject to hydrolysis in the main chain. Therefore, the AmB release depends only on the solvation, the chain relaxation, and the diffusion.

### 3.5. Water Vapor Permeability

A key property that needs to be presented by modern wound dressings is the ability to absorb excessive exudates while maintaining a moist environment [67]. Thus, the water vapor transmission of the PVA–H and PVA–AmB hydrogels was evaluated. The results showed that the presence of AmB in the hydrogel network does not significantly affect the water vapor permeability (PVA–AmB = 393 ± 33 $g \cdot m^{-2} \cdot day^{-1}$; PVA–H = 452 ± 10 $g \cdot m^{-2} \cdot day^{-1}$) ($p$-value > 0.05) (Figure 4A).

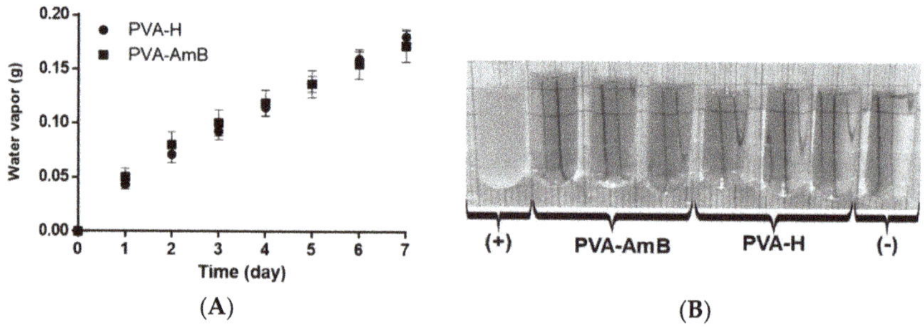


**Figure 4.** (**A**) Water vapor transmission rate of PVA–H and PVA–AmB hydrogels. (**B**) Medium aspect of positive (+) and negative (−) controls, as well as PVA–H and PVA–AmB after 7 days exposure to the environment, showing the resistance of the hydrogels to microbial permeability.

The literature reports that the skin in a healthy state has a water vapor permeability of 204 $g \cdot m^{-2} \cdot day^{-1}$, while in an injured state it can fluctuate between 279 to 5138 $g \cdot m^{-2} \cdot day^{-1}$ [68,69]. Thus, based on our results, it is possible to infer that the PVA–AmB would be able to control the water loss by evaporation from a wound, allowing the skin in an injured state to have a water vapor permeability compatible with that in the healthy state, avoiding dehydration.

### 3.6. Microbial Permeability Assay

The ulcer, the most common clinical manifestation of CL, can remain active for several months, exposing the tissue to microorganisms both from normal skin microbiota and from the environment. The role of the secondary infection in the healing evolution of the leishmaniotic ulcers is still unclear in the literature. However, their acquisition can be considered a grievance to the patient clinical condition [6]. Furthermore, prevention of secondary infection is especially important for *Leishmania*-human immunodeficiency virus (HIV) co-infection individuals, who already have a deficient immunological system [70,71]. Therefore, wound dressing formulations able to treat *Leishmania* infection and avoid secondary microbial infections seem to be of particular importance to these patients.

Thus, the resistance of the hydrogels to microbial permeability was evaluated for 7 days. During the assay, no cloudiness was observed in the culture medium (Figure 4B), which reveals the ability of the hydrogels to act as a barrier for the microorganisms. In fact, this result shows that the use of AmB-loaded hydrogels for treatment of CL may prevent and/or hinder the development of secondary opportunistic infections. As evidenced by the SEM analysis in Figure 5A, the hydrogel displays an apparent nonporous structure. However, it is important to highlight that according to its porosity, the hydrogels are often classified as nonporous/nanoporous (10–100 Å), microporous (100–1000 Å), macroporous (0.1–1 μm) and superporous systems (1–1000 μm) [38,72–74]. Since the magnification of 1500× did not provide enough resolution to draw an accurate conclusion about the porosity of the system on a scale smaller than 1 μm (Figure 5A) and, beyond that, a collapse of the hydrogel structure was caused (Figure 5B), no analysis at higher magnification could be performed. Therefore, such classification should be confirmed by further characterization.

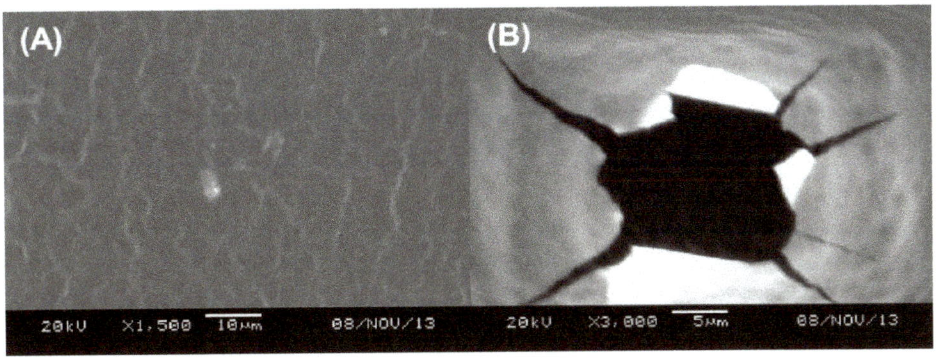


**Figure 5.** Scanning electron microscope image of PVA–AmB hydrogel evidencing the sample deformation, caused by the electron beam, as the magnification increment goes from 1500× (**A**) to 3000× (**B**).

Regardless of its classification, either a nonporous system or a microporous one with pores smaller than the pathogens, the system may act as a mechanical barrier. Therefore, the microbial impermeability of the hydrogels produced in this work makes it potentially more efficient in preventing secondary infection than conventional materials such as gauze. [75].

### 3.7. Biological Activity Assays

The leishmanicidal activity of the system against *L. amazonensis* and *L. braziliensis* promastigotes was evaluated for 48 h in a culture medium containing PVA–H and PVA–AmB hydrogels. The results indicated that AmB released from the system performed a similar cytotoxic pattern for *Leishmania* promastigotes (i.e. 100% and 99% ± 2% of mortality for *L. amazonensis* and *L. braziliensis*, respectively) when compared to the positive control, within the first 24 h, evidenced by the absence of flagellar motility of the parasites (Figure 6A,B). Based on the drug release assay (Section 3.4), this is an expected result, since it was expected that the system would release ~38% ± 4% of the initial dose. Therefore, a concentration of AmB equivalent to the positive control (54 ± 6 μg/mL) would be produced.

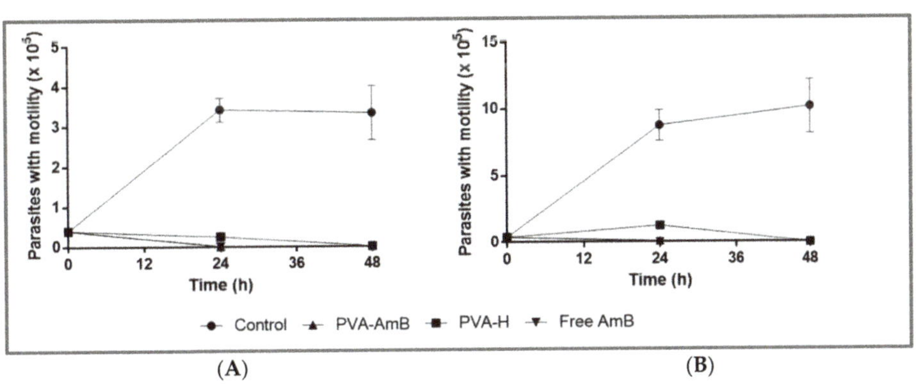


**Figure 6.** Leishmanicidal activity of the PVA–H and the PVA–AmB against (**A**) *Leishmania amazonensis* and (**B**) *Leishmania braziliensis*.

Zauli-Nascimento et al [56] evaluated the AmB susceptibility of both *Leishmania* species using ATCC strains and clinical strains. The authors reported an AmB effective inhibition concentration of 50% (EC50) ranging from 36 ± 4 to 92 ± 4 ng/mL for the *L. braziliensis* and from 55 ± 1 to 83 ± 6 ng/mL for the *L. amazonensis*. In addition, no significant differences were found among *Leishmania* spp. clinical isolate susceptibility, showing a range of AmB EC90 from 80 to 650 ng/mL.

Thus, considering these findings and assuming that the AmB release in our study had followed the same profile observed in the in vitro drug release assay, the amount of AmB released, within the first 24 h (54 ± 6 µg/mL), would be enough to kill more than 90% of the promastigote forms. Furthermore, considering that the susceptibility of the *Leishmania* promastigote and amastigote forms to AmB was similar, as previously reported [76], it is possible to infer that a similar response to the PVA–AmB observed in vitro for *Leishmania* promastigotes would be observed for the amastigote stage.

Regarding the leishmanicidal activity of the PVA–H, an unexpected decrease in the number of parasites was observed when compared to the negative control (Figure 6A,B). This could be explained by the presence of residual cross-linking inside the polymer matrix, decreasing the promastigote viability.

In order to better understand the AmB cytotoxicity in VERO cells an AmB kill curve was carried out. It seems that the cell viability displays a reasonable linear correlation to the concentration of AmB, as demonstrated by fitting the data through linear regression (Figure 7A). Regarding the cell viability of the hydrogels, no significant difference between PVA–H and PVA–AmB ($p > 0.05$) was observed.

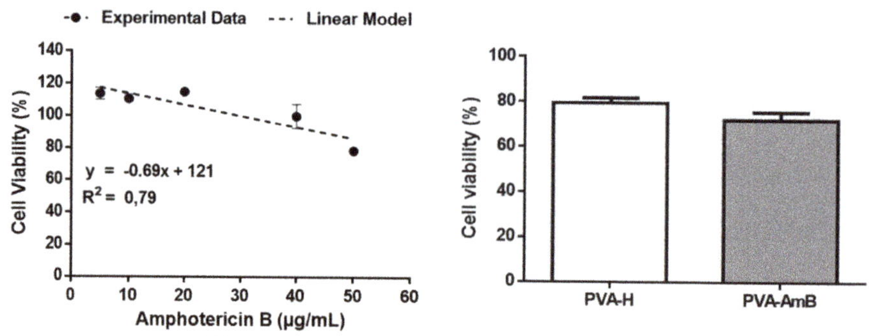


**Figure 7.** AmB killing curve for VERO cell lines (**A**); evaluation of the cytotoxicity of PVA-H and PVA-AmB against VERO cell lines (**B**).

Performing the same assumption as before, which is the pattern of AmB released into the culture medium follows the one previously observed (Section 3.4), it is possible to compare the cell viability obtained from the hydrogels (Figure 7B, PVA–H = 80% ± 5%; PVA–AmB = 72% ± 7%) to the one predicted by the linear model (~102%). Even though these values of toxicity enable to classify the systems as no cytotoxic potential [77], this analysis revealed that the systems displayed a cytotoxicity 20% higher than expected, which might be related to the presence of residual cross-linking agents.

Furthermore, it is important to highlight that the system here proposed is intended for topical use. Therefore, it is expected that a high dilution factor of the molecules will perhaps reach the bloodstream and, thereafter, the kidneys. For instance, even though the whole dose used on the assay with *Leishmania* spp. (~216 µg) reaches the bloodstream, a dilution by a factor of ~6 L is expected [78]. Therefore, obtaining a concentration of AmB ~36 ng/mL. Taking into account that AmB is highly bound in plasma protein >95% [79], only ~5% will be effectively evaluable to perform the cytotoxic effect. However, no toxic effect is expected with AmB concentrations below 40 µg/mL (Figure 7A).

Due to the recognized use of AmB as a gold standard on the treatment of fungal infections, mainly the ones caused by *Candida* spp., the antifungal properties of PVA–AmB hydrogels were evaluated by the radial disc diffusion assay against *C. albicans*. The zone of inhibition was measured using free

AmB and PVA–H as positive and negative controls, respectively. The PVA–AmB hydrogel resulted in a significant reduction in the viability of *C. albicans*, with inhibitory zones on the agar surface bigger than the positive control (16.7 ± 3 mm vs. 9.8 ± 0.5 mm for 24 h and 16.3 ± 2.5 vs. 9.5 ± 1.3 mm for 48 h, respectively, $p < 0.05$). These results indicate that the AmB entrapped inside of the hydrogel structure keeps its antifungal activity, just as free AmB, while the PVA–H did not show any antifungal activity, corroborating with the idea that AmB stands with its pharmacological activity unaltered and the residual cross-linking quantities were negligible to show antifungal activity or severe cytotoxicity. It has been reported that the affinity of AmB to the ergosterol on fungi membranes is strongly dependent on its aggregation state [80,81]. This phenomenon could be the main reason behind the enhanced antifungal activity of PVA–AmB compared to the free form of the drug. In fact, it is expected that the polymeric matrix of the hydrogel will release the AmB slowly and continuously on its monomeric form. On the other hand, free AmB, which is already evaluable into the medium, should display the aggregated form, since its concentration is higher than $10^{-7}$ M [82–84].

## 4. Conclusions

The results presented here show that the factorial design was a useful tool for the development of PVA hydrogels by the casting method. Parameters like pH, cross-linker, and AmB concentration demonstrated significant effects over the system's structure. Importantly, the AmB loaded into PVA hydrogels remained in its amorphous state and did not significantly change the crystallinity index. Besides, the PVA–AmB showed water vapor permeation properties that make this kind of system suitable for topical treatment of CL lesions. Furthermore, the applied mathematical modeling showed that the AmB presence in the system altered the swelling mechanism from less Fickian to anomalous, by increasing the hydrogel swelling capacity. However, no burst release of AmB was observed and the release kinetic profile was adequately fitted to the Higuchi model. Additionally, AmB-loaded hydrogels were resistant to microbial permeation and showed effective activity against *Leishmania* promastigotes and *Candida albicans*, as well as no cytotoxic potential for VERO cell lines. These results are indicative of an efficient pharmacological activity and suitable biocompatibility of PVA–AmB hydrogels and show the great potential application of this system for the topical treatment of CL.

**Author Contributions:** Conceptualization, F.A.Jr. and K.G.d.H.eS; methodology, F.A.Jr.; validation, F.A.Jr and M.C.L.C.F; formal analysis, F.A.Jr. and H.R.M.; investigation, F.A.Jr., M.C.L.C.F, E.A.C. and V.d.O.F.L.; resources, E.S.T.d.E.; writing—original draft preparation, F.A.Jr., H.R.M. and J.G.; writing—review and editing, K.G.d.H.eS, A.G.d.O. and E.S.T.d.E.; visualization E.S.T.d.E. and A.G.d.O.; supervision, E.S.T.d.E. and K.G.d.H.eS; project administration, E.S.T.d.E. and J.G; funding acquisition, E.S.T.d.E.

**Funding:** This research was funded by the Coordenação de Aperfeiçoamento de Pessoal de Nível Superior (CAPES), Finance Code 001 and the Conselho Nacional de Desenvolvimento Científico e Tecnológico (CNPq).

**Acknowledgments:** The authors would like to thank Paulo Henrique de Souza Picciani from Instituto de Macromoléculas Professora Eloisa Mano (IMA, UFRJ), M.Sc. Karen Cybelle de Holanda Silva and PharmB Bartolomeu Santos de Souza from Laboratório de Sistemas Dispersos (LaSiD, UFRN) for the technical support during the development of this study. The authors are also grateful to Glenn Hawes, M Ed. (Master of English Education, University of Georgia) for editing this manuscript.

**Conflicts of Interest:** The authors declare no conflict of interest.

## References

1. Alvar, J.; Arana, B.I. Appraisal of Leishmaniasis Chemotherapy, Current Status and Pipeline StrategiesChapter 1 Leishmaniasis, Impact and Therapeutic Needs. In *Drug Discovery for Leishmaniasis*; The Royal Society of Chemistry: London, UK, 2018; pp. 1–13.
2. WHO. *Sustaining the Drive to Overcome the Global Impact of Neglected Tropical Diseases: Second WHO Report on Neglected Tropical Diseases*; WHO Library: Paris, France, 2013; pp. 67–71.
3. WHO. *Leishmaniasis—Fact sheet N°375*; WHO: Geneva, Switzerland, 2014.
4. Okwor, I.; Uzonna, J. Vaccines and vaccination strategies against human cutaneous leishmaniasis. *Hum. Vaccines* **2009**, *5*, 291–301. [CrossRef]

5. Bacon, K.M.; Hotez, P.J.; Kruchten, S.D.; Kamhawi, S.; Bottazzi, M.E.; Valenzuela, J.G.; Lee, B.Y. The potential economic value of a cutaneous leishmaniasis vaccine in seven endemic countries in the Americas. *Vaccine* **2013**, *31*, 480–486. [CrossRef]
6. WHO. Control of the Leishmaniasis. In Proceedings of the Meeting of the WHO Expert Committee on the Control of Leishmaniases, Geneva, Switzerland, 22–26 March 2010.
7. Neves, L.O.; Talhari, A.C.; Gadelha, E.P.N.; da Silva Júnior, R.M.; Guerra, J.A.O.; Ferreira, L.C.L.; Talhari, S. Estudo clínico randomizado comparando antimoniato de meglumina, pentamidina e anfotericina B para o tratamento da leishmaniose cutânea ocasionada por Leishmania guyanensis. *An. Bras. Dermatol.* **2011**, *86*, 1092–1101. [CrossRef] [PubMed]
8. Hamill, R.J.; Sobel, J.D.; El-Sadr, W.; Johnson, P.C.; Graybill, J.R.; Javaly, K.; Barker, D.E.; Baker, C.J.; AmBisome Cryptococcal Meningitis Study Group. Comparison of 2 Doses of Liposomal Amphotericin B and Conventional Amphotericin B Deoxycholate for Treatment of AIDS-Associated Acute Cryptococcal Meningitis: A Randomized, Double-Blind Clinical Trial of Efficacy and Safety. *Clin. Infect. Dis.* **2010**, *51*, 225–232. [CrossRef] [PubMed]
9. Solomon, M.; Pavlotzky, F.; Barzilai, A.; Schwartz, E. Liposomal amphotericin B in comparison to sodium stibogluconate for Leishmania braziliensis cutaneous leishmaniasis in travelers. *J. Am. Acad. Dermatol.* **2013**, *68*, 284–289. [CrossRef] [PubMed]
10. Mehta, J.; Blake, J.; Craddock, C. Comparative efficacy of amphotericin B lipid complex and liposomal amphotericin B for the treatment of invasive fungal infections in HSCT recipients and other immunocompromised patient populations with hematologic malignancies: A critical review. *Open Transplant. J.* **2011**, *5*, 23–29. [CrossRef]
11. El-Sayed, M.; Anwar, A.E. Intralesional sodium stibogluconate alone or its combination with either intramuscular sodium stibogluconate or oral ketoconazole in the treatment of localized cutaneous leishmaniasis: A comparative study. *J. Eur. Acad. Dermatol. Venereol.* **2009**, *24*, 335–340. [CrossRef]
12. Asilian, A.; Sadeghinia, A.; Faghihi, G.; Momeni, A. Comparative study of the efficacy of combined cryotherapy and intralesional meglumine antimoniate (Glucantime) vs. cryotherapy and intralesional meglumine antimoniate (Glucantime) alone for the treatment of cutaneous leishmaniasis. *Int. J. Pharm.* **2004**, *43*, 281–283. [CrossRef]
13. Safi, N. Evaluation of thermotherapy for the treatment of cutaneous Leishmaniasis in Kabul, Afghanistan: A randomized controlled trial. *Int. J. Infect. Dis.* **2005**, *16* (Suppl. 1), 1148–1155. [CrossRef]
14. Kim, D.H.; Chung, H.J.; Bleys, J.; Ghohestani, R.F. Is paromomycin an effective and safe treatment against cutaneous leishmaniasis? A aeta-analysis of 14 randomized controlled trials. *PLoS Negl. Trop. Dis.* **2009**, *3*, e381. [CrossRef]
15. Frankenburg, S.; Glick, D.; Klaus, S.; Barenholz, Y. Efficacious topical treatment for murine cutaneous leishmaniasis with ethanolic formulations of amphotericin B. *Antimicrob. Agents Chemother.* **1998**, *42*, 3092–3096. [CrossRef] [PubMed]
16. Kaya, A.Z.; Turani, N.; Akyuz, M. The effectiveness of a hydrogel dressing compared with standard management of pressure ulcers. *J. Wound Care* **2005**, *14*, 42–44. [CrossRef] [PubMed]
17. Zhao, X.; Wu, H.; Guo, B.; Dong, R.; Qiu, Y.; Ma, P.X. Antibacterial anti-oxidant electroactive injectable hydrogel as self-healing wound dressing with hemostasis and adhesiveness for cutaneous wound healing. *Biomaterials* **2017**, *122*, 34–47. [CrossRef] [PubMed]
18. Zoellner, P.; Kapp, H.; Smola, H. Clinical performance of a hydrogel dressing in chronic wounds: A prospective observational study. *J. Wound Care* **2007**, *16*, 133–136. [CrossRef]
19. Singh, B.; Pal, L. Sterculia crosslinked PVA and PVA-poly(AAm) hydrogel wound dressings for slow drug delivery: Mechanical, mucoadhesive, biocompatible and permeability properties. *J. Mech. Behav. Biomed.* **2012**, *9*, 9–21. [CrossRef] [PubMed]
20. Yan, J.; Chen, X.; Yu, S.; Zhou, H. Comparison of different in vitro mucoadhesion testing methods for hydrogels. *J. Drug Deliv. Sci. Technol.* **2017**, *40*, 157–163. [CrossRef]
21. Han, F.; Dong, Y.; Song, A.; Yin, R.; Li, S. Alginate/chitosan based bi-layer composite membrane as potential sustained-release wound dressing containing ciprofloxacin hydrochloride. *Appl. Surf. Sci.* **2014**, *311*, 626–634. [CrossRef]

22. Ci, L.-Q.; Huang, Z.-G.; Lv, F.-M.; Wang, J.; Feng, L.-L.; Sun, F.; Cao, S.-J.; Liu, Z.-P.; Liu, Y.; Wei, G.; et al. Enhanced Delivery of Imatinib into Vaginal Mucosa via a New Positively Charged Nanocrystal-Loaded in Situ Hydrogel Formulation for Treatment of Cervical Cancer. *Pharmaceutics* **2019**, *11*, 15. [CrossRef]
23. Jøraholmen, M.W.; Basnet, P.; Tostrup, M.J.; Moueffaq, S.; Škalko-Basnet, N. Localized Therapy of Vaginal Infections and Inflammation: Liposomes-In-Hydrogel Delivery System for Polyphenols. *Pharmaceutics* **2019**, *11*, 53. [CrossRef]
24. Mateus, D.; Marto, J.; Trindade, P.; Gonçalves, H.; Salgado, A.; Machado, P.; Melo-Gouveia, A.; Ribeiro, H.M.; Almeida, A.J. Improved Morphine-Loaded Hydrogels for Wound-Related Pain Relief. *Pharmaceutics* **2019**, *11*, 76. [CrossRef]
25. Wichterle, O.; LÍM, D. Hydrophilic gels for biological use. *Nature* **1960**, *185*, 117–118. [CrossRef]
26. Jiang, S.; Liu, S.; Feng, W.H. PVA hydrogel properties for biomedical application. *J. Mech. Behav. Biomed. Mater.* **2011**, *4*, 1228–1233. [CrossRef] [PubMed]
27. Dou, X.; Li, P.; Schonherr, H. Three-Dimensional Microstructured Poly(vinyl alcohol) Hydrogel Platform for the Controlled Formation of Multicellular Cell Spheroids. *Biomacromolecules* **2018**, *19*, 158–166. [CrossRef] [PubMed]
28. Kobayashi, M.; Chang, Y.S.; Oka, M. A two year in vivo study of polyvinyl alcohol-hydrogel (PVA-H) artificial meniscus. *Biomaterials* **2005**, *26*, 3243–3248. [CrossRef] [PubMed]
29. Wan, W.K.; Campbell, G.; Zhang, Z.F.; Hui, A.J.; Boughner, D.R. Optimizing the tensile properties of polyvinyl alcohol hydrogel for the construction of a bioprosthetic heart valve stent. *J. Biomed. Mater. Res.* **2002**, *63*, 854–861. [CrossRef] [PubMed]
30. Pillai, M.M.; Gopinathan, J.; Kumar, R.S.; Kumar, G.S.; Shanthakumari, S.; Sahanand, K.S.; Bhattacharyya, A.; Selvakumar, R. Tissue engineering of human knee meniscus using functionalized and reinforced silk-polyvinyl alcohol composite three-dimensional scaffolds: Understanding the in vitro and in vivo behavior. *J. Biomed. Mater. Res. Part A* **2018**, *106*, 1722–1731. [CrossRef] [PubMed]
31. Matsumura, K.; Hayami, T.; Hyon, S.H.; Tsutsumi, S. Control of proliferation and differentiation of osteoblasts on apatite-coated poly(vinyl alcohol) hydrogel as an artificial articular cartilage material. *J.Biomed. Mater. Res. Part A* **2010**, *92*, 1225–1232. [CrossRef] [PubMed]
32. Gao, D.Q.; Zhou, X.Q.; Gao, Z.H.; Shi, X.C.; Wang, Z.L.; Wang, Y.; Zhang, P.B. Preparation and Characterization of Silver Sulfadiazine-Loaded Polyvinyl Alcohol Hydrogels as an Antibacterial Wound Dressing. *J. Pharm. Sci.* **2018**, *107*, 2377–2384. [CrossRef]
33. Gao, T.L.; Jiang, M.H.; Liu, X.Q.; You, G.J.; Wang, W.Y.; Sun, Z.H.; Ma, A.G.; Chen, J. Patterned Polyvinyl Alcohol Hydrogel Dressings with Stem Cells Seeded for Wound Healing. *Polymers* **2019**, *11*, 171. [CrossRef] [PubMed]
34. Hyon, S.H.; Cha, W.I.; Ikada, Y.; Kita, M.; Ogura, Y.; Honda, Y. Poly(vinyl alcohol) hydrogels as soft contact-lens material. *J. Biomater. Sci. Polym. Ed.* **1994**, *5*, 397–406. [CrossRef]
35. Freire, M.C.L.C.; Alexandrino, F.; Marcelino, H.R.; Picciani, P.H.d.S.; Silva, K.G.d.H.e.; Genre, J.; Oliveira, A.G.d.; Egito, E.S.T.d. Understanding drug release data through thermodynamic analysis. *Materials* **2017**, *10*, 651. [CrossRef]
36. Hickey, A.S.; Peppas, N.A. Mesh size and diffusive characteristics of semicrystalline poly(vinyl alcohol) membranes prepared by freezing/thawing techniques. *J. Membr. Sci.* **1995**, *107*, 229–237. [CrossRef]
37. Li, T.; Lei, Y.; Guo, M.; Yan, H. Crosslinked poly(vinyl alcohol) hydrogel microspheres containing dispersed fenofibrate nanocrystals as an oral sustained delivery system. *Eur. Polym. J.* **2018**, *101*, 77–82. [CrossRef]
38. Peppas, N.A.; Bures, P.; Leobandung, W.; Ichikawa, H. Hydrogels in pharmaceutical formulations. *Eur. J. Pharm. Biopharm.* **2000**, *50*, 27–46. [CrossRef]
39. Qi, M.; Gu, Y.; Sakata, N.; Kim, D.; Shirouzu, Y.; Yamamoto, C.; Hiura, A.; Sumi, S.; Inoue, K. Pva hydrogel sheet macroencapsulation for the bioartificial pancreas. *Biomaterials* **2004**, *25*, 5885–5892. [CrossRef]
40. Nakano, K.; Tozuka, Y.; Takeuchi, H. Effect of surface properties of liposomes coated with a modified polyvinyl alcohol (PVA-R) on the interaction with macrophage cells. *Int. J. Pharm.* **2008**, *354*, 174–179. [CrossRef] [PubMed]
41. Suktham, K.; Koobkokkruad, T.; Saesoo, S.; Saengkrit, N.; Surassmo, S. Physical and biological characterization of sericin-loaded copolymer liposomes stabilized by polyvinyl alcohol. *Colloids Surf. B Biointerfaces* **2016**, *148*, 487–495. [CrossRef]

42. Alves, M.H.; Young, C.J.; Bozzetto, K.; Poole-Warren, L.A.; Martens, P.J. Degradable, click poly(vinyl alcohol) hydrogels: Characterization of degradation and cellular compatibility. *Biomed. Mater.* **2012**, *7*, 024106. [CrossRef] [PubMed]
43. DeMerlis, C.C.; Schoneker, D.R. Review of the oral toxicity of polyvinyl alcohol (PVA). *Food Chem. Toxicol.* **2003**, *41*, 319–326. [CrossRef]
44. Mansur, H.S.; Sadahira, C.M.; Souza, A.N.; Mansur, A.A.P. FTIR spectroscopy characterization of poly (vinyl alcohol) hydrogel with different hydrolysis degree and chemically crosslinked with glutaraldehyde. *Mater. Sci. Engi. C Biomim. Supramol. Syst.* **2008**, *28*, 539–548. [CrossRef]
45. Sittiwong, J.; Niamlang, S.; Paradee, N.; Sirivat, A. Electric field-controlled benzoic acid and sulphanilamide delivery from poly(vinyl alcohol) hydrogel. *AAPS PharmSciTech* **2012**, *13*, 1407–1415. [CrossRef]
46. Tripathi, R.; Mishra, B. Development and evaluation of sodium alginate-polyacrylamide graft-co-polymer-based stomach targeted hydrogels of famotidine. *AAPS PharmSciTech* **2012**, *13*, 1091–1102. [CrossRef]
47. *Farmacopeia Brasileira*; ANVISA (Ed.) FioCruz: Brasília, Brazil, 2010; Volume 1, p. 510.
48. Veeruraj, A.; Liu, L.; Zheng, J.; Wu, J.; Arumugam, M. Evaluation of astaxanthin incorporated collagen film developed from the outer skin waste of squid Doryteuthis singhalensis for wound healing and tissue regenerative applications. *Mater. Sci. Eng. C* **2019**, *95*, 29–42. [CrossRef]
49. Lemke, A.; Kiderlen, A.F.; Kayser, O. Amphotericin B. *Appl. Microbiol. Biotechnol.* **2005**, *68*, 151–162. [CrossRef]
50. Santos, C.M.; Oliveira, R.B.; Arantes, V.T.; Caldeira, L.R.; Oliveira, M.C.; Egito, E.S.T.; Ferreira, L.A.M. Amphotericin B-Loaded Nanocarriers for Topical Treatment of Cutaneous Leishmaniasis: Development, Characterization, and In Vitro Skin Permeation Studies. *J. Biomed. Nanotechnol.* **2012**, *8*, 322–329. [CrossRef]
51. Zhang, Y.; Huo, M.; Zhou, J.; Zou, A.; Li, W.; Yao, C.; Xie, S. DDSolver: An add-in program for modeling and comparison of drug dissolution profiles. *AAPS J.* **2010**, *12*, 263–271. [CrossRef]
52. ASTM. Standard Test Methods for Water Vapor Transmission of Materials. In *Annual Book of ASTM Standards*; ASTM International: West Conshohocken, PA, USA, 1996.
53. Vargas, E.A.T.; do Vale Baracho, N.C.; de Brito, J.; de Queiroz, A.A.A. Hyperbranched polyglycerol electrospun nanofibers for wound dressing applications. *Acta Biomater.* **2010**, *6*, 1069–1078. [CrossRef]
54. Perez, A.P.; Altube, M.J.; Schilrreff, P.; Apezteguia, G.; Celes, F.S.; Zacchino, S.; de Oliveira, C.I.; Romero, E.L.; Morilla, M.J. Topical amphotericin B in ultradeformable liposomes: Formulation, skin penetration study, antifungal and antileishmanial activity in vitro. *Colloids Surf. B Biointerfaces* **2016**, *139*, 190–198. [CrossRef]
55. Ruiz, H.K.; Serrano, D.R.; Dea-Ayuela, M.A.; Bilbao-Ramos, P.E.; Bolás-Fernández, F.; Torrado, J.J.; Molero, G. New amphotericin B-gamma cyclodextrin formulation for topical use with synergistic activity against diverse fungal species and *Leishmania* spp. *Int. J. Pharm.* **2014**, *473*, 148–157. [CrossRef]
56. Zauli-Nascimento, R.C.; Miguel, D.C.; Yokoyama-Yasunaka, J.K.; Pereira, L.I.; Pelli de Oliveira, M.A.; Ribeiro-Dias, F.; Dorta, M.L.; Uliana, S.R. In vitro sensitivity of *Leishmania* (*Viannia*) braziliensis and *Leishmania* (*Leishmania*) *amazonensis* Brazilian isolates to meglumine antimoniate and amphotericin B. *Trop. Med. Int. Health* **2010**, *15*, 68–76. [CrossRef]
57. Falk, R.; Domb, A.J.; Polacheck, I. A novel injectable water-soluble amphotericin B-arabinogalactan conjugate. *Antimicrob. Agents Chemother.* **1999**, *43*, 1975–1981. [CrossRef]
58. Reis, E.F.; Campos, F.S.; Lage, A.p.; Leite, R.C.; Heneine, L.G.; Vasconcelos, W.L.; Lobato, Z.I.P.; Mansur, H.S. Synthesis and characterization of poly (vinyl alcohol) hydrogels and hybrids for rMPB70 protein adsorption. *Mater. Res.* **2006**, *9*, 185–191. [CrossRef]
59. Clarke, E.G.C. Amphotericin. In *Clarke's Isolation and Identification of Drugs in Pharmaceuticals, Body Fluids and Post-Mortem Material*, 2nd ed.; Moffat, A.C., Ed.; Pharmaceutical Press: London, 1986; p. 1200.
60. Chang, M.; Liu, X.; Meng, L.; Wang, X.; Ren, J. Xylan-Based Hydrogels as a Potential Carrier for Drug Delivery: Effect of Pore-Forming Agents. *Pharmaceutics* **2018**, *10*, 261. [CrossRef] [PubMed]
61. Lin, C.-C.; Metters, A.T. Hydrogels in controlled release formulations: Network design and mathematical modeling. *Adv. Drug Deliv. Rev.* **2006**, *58*, 1379–1408. [CrossRef] [PubMed]
62. Coughlan, D.C.; Quilty, F.P.; Corrigan, O.I. Effect of drug physicochemical properties on swelling/deswelling kinetics and pulsatile drug release from thermoresponsive poly(*N*-isopropylacrylamide) hydrogels. *J. Control. Release* **2004**, *98*, 97–114. [CrossRef]

63. Ruiz, J.; Mantecón, A.; Cádiz, V. Synthesis and properties of hydrogels from poly (vinyl alcohol) and ethylenediaminetetraacetic dianhydride. *Polymer* **2001**, *42*, 6347–6354. [CrossRef]
64. Vera, L.A.; Macedo, J.L.; Ciuffo, I.A.; Santos, C.G.; Santos, J.B. Antimicrobial susceptibility of aerobic bacteria isolated from leishmaniotic ulcers in Corte de Pedra, BA. *Rev. Soc. Bras. Med. Trop.* **2006**, *39*, 47–50. [CrossRef]
65. Ritger, P.L.; Peppas, N.A. A simple equation for description of solute release II. Fickian and anomalous release from swellable devices. *J. Control. Release* **1987**, *5*, 37–42. [CrossRef]
66. Gu, B.; Sun, X.; Papadimitrakopoulos, F.; Burgess, D.J. Seeing is believing, PLGA microsphere degradation revealed in PLGA microsphere/PVA hydrogel composites. *J. Control. Release* **2016**, *228*, 170–178. [CrossRef] [PubMed]
67. Sibbald, R.G.; Orsted, H.; Schultz, G.S.; Coutts, P.; Keast, D.; Board, I.W.B.P.A.; Board, C.C.W.A. Preparing the wound bed 2003: Focus on infection and inflammation. *Ostomy Wound Manag.* **2003**, *49*, 24–51.
68. Gu, S.Y.; Wang, Z.M.; Ren, J.; Zhang, C.Y. Electrospinning of gelatin and gelatin/poly(L-lactide) blend and its characteristics for wound dressing. *Mater. Sci. Eng. C* **2009**, *29*, 1822–1828. [CrossRef]
69. Mi, F.L.; Shyu, S.S.; Wu, Y.B.; Lee, S.T.; Shyong, J.Y.; Huang, R.N. Fabrication and characterization of a sponge-like asymmetric chitosan membrane as a wound dressing. *Biomaterials* **2001**, *22*, 165–173. [CrossRef]
70. Ngouateu, O.B.; Kollo, P.; Ravel, C.; Dereure, J.; Kamtchouing, P.; Same-Ekobo, A.; von Stebut, E.; Maurer, M.; Dondji, B. Clinical features and epidemiology of cutaneous leishmaniasis and Leishmania major/HIV co-infection in Cameroon: Results of a large cross-sectional study. *Trans. R. Soc. Trop. Med. Hyg.* **2012**, *106*, 137–142. [CrossRef]
71. Carranza-Tamayo, C.O.; de Assis, T.S.; Neri, A.T.; Cupolillo, E.; Rabello, A.; Romero, G.A. Prevalence of Leishmania infection in adult HIV/AIDS patients treated in a tertiary-level care center in Brasilia, Federal District, Brazil. *Trans. R. Soc. Trop. Med. Hyg.* **2009**, *103*, 743–748. [CrossRef]
72. Peppas, N.A.; Hilt, J.Z.; Khademhosseini, A.; Langer, R. Hydrogels in Biology and Medicine: From Molecular Principles to Bionanotechnology. *Adv. Mater.* **2006**, *18*, 1345–1360. [CrossRef]
73. Kim, S.; Kim, J.-H.; Jeon, O.; Kwon, I.C.; Park, K. Engineered polymers for advanced drug delivery. *Eur. J. Pharm. Biopharm.* **2009**, *71*, 420–430. [CrossRef]
74. Garg, T.; Singh, S.; Goyal, A.K. Stimuli-Sensitive Hydrogels: An Excellent Carrier for Drug and Cell Delivery. *Crit. Rev.* **2013**, *30*, 369–409. [CrossRef]
75. Baranoski, S.; Ayello, E.A. Wound Treatment Options. In *Wound Care Essentials: Practice Principles*, 3rd ed.; Lippincott Williams & Wilkins: Philadelphia, PA, USA, 2012; p. 183.
76. Escobar, P.; Matu, S.; Marques, C.; Croft, S.L. Sensitivities of Leishmania species to hexadecylphosphocholine (miltefosine), ET-18-OCH(3) (edelfosine) and amphotericin B. *Acta Trop.* **2002**, *81*, 151–157. [CrossRef]
77. ISO. *Biological Evaluation of Medical Devices—Part 5: Tests for in Vitrocytotoxicity*, 3rd ed.; International Organization for Standardization: Geneva, Switzerland, 2009; Volume 10993.
78. Davy, K.P.; Seals, D.R. Total blood volume in healthy young and older men. *J. Appl. Physiol.* **1994**, *76*, 2059–2062. [CrossRef]
79. Bekersky, I.; Fielding, R.M.; Dressler, D.E.; Lee, J.W.; Buell, D.N.; Walsh, T.J. Plasma protein binding of amphotericin B and pharmacokinetics of bound versus unbound amphotericin B after administration of intravenous liposomal amphotericin B (AmBisome) and amphotericin B deoxycholate. *Antimicrob. Agents Chemother.* **2002**, *46*, 834–840. [CrossRef]
80. da Silva-Filho, M.A.; da Silva Siqueira, S.D.V.; Freire, L.B.; de Araujo, I.B.; e Silva, K.G.D.H.; da Cunha Medeiros, A.; Araujo-Filho, I.; de Oliveira, A.G.; do Egito, E.S.T. How can micelle systems be rebuilt by a heating process? *Int. J. Nanomed.* **2012**, *7*, 141–150. [CrossRef]
81. Vertut-Croquin, A.; Bolard, J.; Chabbert, M.; Gary-Bobo, C. Differences in the interaction of the polyene antibiotic amphotericin B with cholesterol- or ergosterol-containing phospholipid vesicles. A circular dichroism and permeability study. *Biochemistry* **1983**, *22*, 2939–2944. [CrossRef]
82. Barwicz, J.; Christian, S.; Gruda, I. Effects of the aggregation state of amphotericin B on its toxicity to mice. *Antimicrob. Agents Chemother.* **1992**, *36*, 2310–2315. [CrossRef] [PubMed]

83. Mazerski, J.; Bolard, J.; Borowski, E. Self-association of some polyene macrolide antibiotics in aqueous media. *Biochim. Biophys. Acta (BBA) Gen. Subj.* **1982**, *719*, 11–17. [CrossRef]
84. Tancréde, P.; Barwicz, J.; Jutras, S.; Gruda, I. The effect of surfactants on the aggregation state of amphotericin B. *Biochim. Biophys. Acta (BBA) Biomembr.* **1990**, *1030*, 289–295. [CrossRef]

*pharmaceutics*

MDPI

*Article*

# Thermoreversible Gel-Loaded Amphotericin B for the Treatment of Dermal and Vaginal Candidiasis

**Lilian Sosa [1], Ana Cristina Calpena [1,2], Marcelle Silva-Abreu [1,2,*], Lupe Carolina Espinoza [1,3], María Rincón [1], Nuria Bozal [4], Oscar Domenech [1], María José Rodríguez-Lagunas [5,6] and Beatriz Clares [2,7]**

1 Department of Pharmacy and Pharmaceutical Technology and Physical Chemistry, School of Pharmacy and Food Sciences, University of Barcelona, 08028 Barcelona, Spain
2 Institute de Nanoscience and Nanotechnology (IN2UB), University of Barcelona, 08028 Barcelona, Spain
3 Departamento de Química y Ciencias Exactas, Universidad Técnica Particular de Loja, Loja 1101608, Ecuador
4 Department of Biology, Healthcare and the Environment, School of Pharmacy and Food Sciences, University of Barcelona, 08028 Barcelona, Spain
5 Department of Biochemistry and Physiology, Faculty of Pharmacy and Food Sciences, University of Barcelona, 08028 Barcelona, Spain
6 Institut de Recerca en Nutrició i Seguretat Alimentària (INSA), Universitat de Barcelona, 08921 Barcelona, Spain
7 Department of Pharmacy and Pharmaceutical Technology, School of Pharmacy, University of Granada, 18071 Granada, Spain
* Correspondence: marcellesabreu@gmail.com

Received: 22 May 2019; Accepted: 1 July 2019; Published: 3 July 2019

**Abstract:** The present study was designed to develop a thermoreversible gel of Pluronic (P407) loaded amphotericin B (AmB-gel) for the dermal and vaginal treatment of candidiasis. P407 was used as a copolymer to exploit potential advantages related to increasing drug concentration in the tissue layer in order to provide a local effect. Parameters including internal structure, swelling, porosity, and short-term stability were determined. In addition, drug release profile and ex vivo skin and vaginal permeation studies were carried out. Antifungal efficacy was evaluated against strains of *Candida* spp. and atomic force microscopy (AFM) supported the results. The tolerance of AmB-gel was studied by evaluating biomechanical properties of skin and determining the irritation level in scarified rabbit skin supported by histological analysis. Results confirmed the development of a thermoreversible AmB-gel with high porosity exhibiting Newtonian behavior at 4 °C and pseudoplasticity at 32 °C as well as optimal stability for at least 90 days. The Amb-gel provided a sustained drug release following a Boltzmann sigmoidal model. Non permeation was observed in skin and vaginal mucosa, showing a high retained amount of AmB of 960.0 and 737.3 μg/g/cm$^2$, respectively. In vitro antifungal efficacy showed that AmB-gel was more effective than Free-AmB in inhibiting strains of *Candida* spp. and these results were corroborated by AFM. Finally, tolerance studies showed that its application did not induce skin irritation nor alter its biophysical properties. Together, these results confirmed that AmB-gel could be proposed as a promising candidate for the clinical status in the treatment of skin and vaginal candidiasis.

**Keywords:** thermoreversible gel; poloxamer 407; candidiasis; amphotericin B; skin and vaginal mucosa

## 1. Introduction

Cutaneous fungal infections are a significant cause of morbidity and constitute a critical health issue on a global scale. The most identified fungal pathogenesis that affects people worldwide is based on superficial skin, vaginal and nail infections, with approximately 1.7 billion individuals affected. This has led to an increase of superficial mycoses higher than 20% in recent decades, especially in patients who are immunocompromised or hospitalized [1,2] as well as in cases of topical burn wound infections [3]. Among these fungal infections, various species of Candida cause superficial complications known as candidiasis [4]. Categorically, *Candida albicans* is reported to be the fourth most frequent cause of infection [5].

The first-line drugs used for the treatment of these mycoses are the Azoles antifungal compounds such as fluconazole, ketoconazole, voriconazole, itraconazole, clotrimazole, among others [6,7]. However, mechanisms of resistance to these antifungals have been detected [8,9]. The appearance of these resistance mechanisms leads to the use of other antifungals such as amphotericin B (AmB), especially in immunosuppressed patients [10,11].

AmB is a polyene macrolide of a broad-spectrum with high activity against the most frequently occurring fungi involved in cutaneous mycoses, including Candida [12]. It is considered the gold standard in the treatment of fungal infections and has been found to be effective against azole-resistant fungi due to the creation of transmembrane channels by complexation with the membrane sterols which results in fungal cell death. However, it is not without side effects in systemic administration due to the self-assembly process into higher-order molecular forms [13]. From a physicochemical point of view, AmB is a poorly hydrosoluble, amphoteric, amphiphilic molecule and is difficult to solubilize in organic solvents. These properties block an optimal permeation of the drug through the stratum corneum (SC), which is the major barrier that limits permeation of external substances. For this reason, there is no topical formulation of AmB commercially available at the moment [14]. This fact evidences the need to develop new formulations using excipients with permeation-enhancing properties in order to facilitate the penetration of drug into SC and its distribution from SC to epidermis and dermis. Therefore, the research of new therapeutic approaches for the treatment of superficial mycoses, including dermal and vaginal administration from galenical perspective, is a challenging yet necessary undertaking.

The topical administration of antimicrobials for superficial mycoses offers several advantages, particularly for direct application at the site of infection. Among these are the prevention of systemic toxicity of the drug, effectiveness at an eminently local level and insurance that a sufficient amount of drug is retained in the skin [15].

With the aim of achieving an improved release of AmB, this drug was incorporated to Poloxamer 407 (P407), which is a copolymer of amphiphilic nature consisting of a central hydrophobic block of polypropylene oxide (PPO) flanked by hydrophilic polyethylene (PEG) blocks (PEG–PPO–PEG). The P407 prepared above >18% shows an aqueous solution (4–5 °C) that turns into a gel ~32 °C, making it an ideal candidate for thermoreversible delivery [16]. P407 is a low toxicity excipient approved by the Food and Drug Administration (FDA) [17]. Moreover, it has beneficial properties that help to promote and improve drug permeation through the skin and mucosa [18].

Therefore, an approach to deliver AmB formulated into P407 gel form specifically developed for skin and vaginal treatment against *Candida* spp. was developed. This work has accomplished detailed research of (i) development and physicochemical characterization of AmB-gel for topical administration, (ii) biopharmaceutical studies (release profile) and ex vivo permeation studies, (iii) in vitro antifungal activity and its effect on yeast cells by atomic force microscopy (AFM), and (iv) biomechanical properties and tolerance assay complemented with histological studies.

## 2. Materials and Methods

### 2.1. Materials

AmB was obtained from Acofarma (Barcelona, Spain). Pluronic®® F127 (P407) was obtained from Fagron (Barcelona, Spain). Dimethyl sulfoxide (DMSO), methanol, castor oil and *N*,*N*-dimethylformamide were obtained from Sigma-Aldrich (Darmstadt, Germany). Transcutol®® P and Propylene glycol were kindly provided by Gattefossé (Barcelona, Spain). The water used in all experiments was obtained from a Milli-Q®® Plus System (Millipore Co., Burlington, MA, USA). All the chemicals and reagents were of analytical grade.

### 2.2. *Preparation of AmB-Gel*

Firstly, a selection of solvents was tested to determine the solubility of AmB. Transcutol®® P, Propylene glycol, *N,N*-dimethyformamide, Castor oil and DMSO were analyzed. The DMSO was selected as the best solubilizing component of AmB. Shortly afterwards 25 g of P407 were dispersed on ultrapure water at 4 °C up to 100 mL. Subsequently, 60 mg of AmB were dissolved in 10 mL of DMSO, of which 1 mL was slowly added under stirring to 20 mL of P407 solution at 4 °C. The resulting solution was adjusted to pH 6.5 with NaOH 2N solution, thus obtaining an AmB-gel at a final concentration of 0.03%.

### 2.3. *Physiochemical Characterization*

To analyze the morphology, the AmB-gel was dried over a period of 8 days using a desiccator with provision for a vacuum. Once it was dried, a small quantity was coated with carbon as a conductor agent. The internal structure of the gel was examined by scanning electron microscopy (SEM) using a JEOL J-7100F (Peabody, Massachusetts, MA, USA).

#### 2.3.1. Swelling and Degradation Tests

The swelling ratio (SR) and degradation as percentage of weight loss (WL) were accomplished by a gravimetric method. Half a gram of dried AmB-gel or fresh AmB-gel was used to carry out swelling and degradation tests, respectively. In both experiments the sample was incubated in PBS (pH = 5.5) at 32 °C for 30 min. Samples ($n$ = 3) were removed and weighed after blotting the surface water at predetermined time intervals of 5 min. The PBS uptake was carried out in triplicate.

The SR ratio was calculated using the following equation and expressed by kinetic modeling:

$$\text{SR} = \frac{W_s - W_d}{W_d} \tag{1}$$

where $W_s$ is the weight of the swollen AmB-gel at 5 min intervals and $W_d$ is the weight of dried gel.

WL was calculated following the equation and expressed by kinetic modeling.

$$\text{WL } (\%) = \frac{W_i - W_d}{W_i} \times 100 \tag{2}$$

where $W_i$ is the initial weight of AmB-gel and $W_d$ the weight of gel at different times.

#### 2.3.2. Porosity Study

The porosity percentage (P) was calculated by displacement of the solvent. The method consisted of immersing the previously dried AmB-gel in absolute ethanol for 2 min and then weighing after the excess ethanol on the surface was blotted. The porosity percentage was calculated using the following equation.

$$\text{P} = \frac{w2 - w1}{\rho \times V} \times 100 \tag{3}$$

where, $w1$ represents the weight of the dried AmB-gel to be immersed in ethanol, $w2$ represents the weight of AmB-gel after being immersed in ethanol, ρ is the density of absolute ethanol, and $V$ is the volume of the gel.

### 2.4. *Stability Study*

The AmB-gel was stored at two temperatures: 4 °C and 25 °C. The pH values of formulation were measured in a Crison 501 digital pH/mV-meter (Crison Instruments, Barcelona, Spain) for a period of

90 days (1, 30, 60, and 90 days) at both temperatures. Values were reported as the mean ± standard deviation (SD) of six replicates.

For the chemical stability studies, the formulation was suspended in *N,N*–dimethylformamide: methanol:water (55:15:30, *v*/*v*/*v*) and the amount of drug was quantified by a previously validated method of High Performance Liquid Chromatography (HPLC) (Waters, Milford, MA, USA) [12].

The short-term stability of AmB-gel was studied for a period of 90 days at 4 °C and 25 °C. A TurbiScan Lab®® (Formulaction Co., L'Union, France) was used to analyze the destabilization phenomena by transmission and light retrodispersion using a pulsed near-infrared light source ($\gamma$ = 880 nm) at 25 °C over a span of 90 days.

### 2.5. Rheological Studies

The rheological measurements were performed using a Thermo Scientific Haake Rheostress 1 rotational rheometer (Thermo Fisher Scientific, Kalsruhe, Germany) equipped with cone plate geometry (60 mm diameter, 2° angle) with mobile upper cone Haake C60/2° Ti (0.105 mm gap). Viscosity and flow curves were rehearsed at 4 °C and 32 °C in triplicate. The shear rate ramp program included: 0→50 $s^{-1}$ (3 min), 50 $s^{-1}$ (1 min), and 50→0 $s^{-1}$ (3 min). Obtained data were then fitted to different mathematical models: Newton, Bingham, Ostwald-de Waele, Cross, Casson, and Herschel–Bulkley.

### 2.6. Gelation Time and Spreadability Test

Ten milliliters of AmB-gel was added to a transparent vial with a magnetic bar and placed in a low temperature water bath. The solution was then heated to 37 ± 0.1 °C while being stirred (400 rpm). The gelation time was measured in triplicate once the magnetic bar stopped moving due to gelation.

The spreadability of AmB-gel was determined in triplicate as follows; 0.5 g of formulation was placed within a 1 cm diameter circle previously marked on a glass plate, after which a second glass plate was subsequently placed without sliding. A series of weights (15, 22, 27, 29, 32, and 37 g) were successively added and allowed to rest for 2 min each at 4 °C. The same operation was repeated at 32 °C but using different weights (100, 200, and 300 g). The diameters (cm) of the circle spreads were measured and recorded as comparative values. Experimental data were then fitted to mathematical models using GraphPad Prism®® version 6.0 (GraphPad Software Inc., San Diego, CA, USA).

### 2.7. In Vitro Release and Kinetic Evaluation

In vitro release studies were performed in vertical diffusion Franz cells (FDC-400, Vidra-Foc, Barcelona, Spain) using nylon membranes.

DMSO solution continuously stirred at 32 ± 1 °C was used as the receptor medium to accomplish sink conditions. An aliquot of 1.3 mL of the AmB-gel was placed in the donor phase. Aliquots of 300 µL were collected from the receptor compartment at different times for 24 h and replaced with an equal volume of tempered DMSO kept under stirring at 600 rpm. AmB was quantified by previously validated HPLC [12]. Results are reported as the mean ± SD of six replicates.

Experimental data were fitted to five kinetic models (zero-order, first-order, Peppas–Korsmeyer, Higuchi, and Weibull function) by nonlinear least squares regression using GraphPad Prism®® version 5.01. The best fit was selected based on the Akaike's Information Criterion (AIC) and coefficient determination ($r^2$).

### 2.8. Ex Vivo Permeation Studies: Skin and Vaginal Mucosa

Ex vivo permeation studies were performed as described in Section 2.7 using skin samples with thicknesses of 400 µm or Vaginal porcine mucosae as membranes. Transcutol P®® was used as the receptor medium which was kept at 32 ± 0.5 °C (Human skin) and 37 ± 1 °C (Vaginal porcine mucosa). The experiment was performed with a diffusion area of 0.64 $cm^2$. The human skin was obtained from the abdominal region of a healthy woman after abdominal plastic surgery. The experimental protocol was approved by the Bioethics Committee of Barcelona SCIAS Hospital (Barcelona, Spain) (ref: BEC/001/16) and written informed consent was provided by the volunteer. Integrity of skin was evaluated by measuring the transepidermal water loss (TEWL) with a DermaLab® module (Cortex Technology, Hadsund, Denmark), exhibiting values below 10 g/m$^2$·h.

Vaginal porcine mucosae were obtained under veterinary supervision, from three- to four-month-old pigs after sacrifice using an overdose of sodium thiopental at the Animal Facility at Bellvitge Campus of Barcelona University (Barcelona, Spain) in accordance with protocols prescribed by the Animal Experimentation Ethics Committee of the University of Barcelona, Spain (CEEA-UB). Mucosa was placed into of Hank's balanced salt solution (HBSS) and refrigerated until use. The experiments were carried out for 36 h (Human skin) and 6 h (Vaginal porcine mucosae), respectively.

At the end of the permeation study, the human skin and porcine vaginal mucosa were removed from the Franz diffusion cell, cleaned with a dodecyl sulfate solution 0.05% and washed in distilled water. AmB retained in the skin and vaginal mucosa was extracted with DMSO for a period of 20 min under cold sonication in an ultrasound bath. The resulting samples were measured by HPLC.

### 2.9. *Antifungal Efficacy*

The minimal inhibitory concentration (MIC), defined as the lowest concentration of an antimicrobial agent that inhibits the growth of a microorganism, was calculated by the broth microdilution method against *C. albicans* ATCC 10231, *C. glabrata* ATCC 66,032, and *C. parapsilosis* ATCC 22,019 strains (American Type Culture Collection, Manassas, VA, USA) following the procedure outlined by the European Committee for Antimicrobial Susceptibility Testing Guidelines (EUCAST) [19] and the Reference method CLSI M27-A3 [20].

This Standards Method provides a valid procedure for testing the susceptibility of glucose-fermenting yeasts to antifungal agents by determining the MIC. A synthetic medium containing RPMI-1640, glutamine, pH indicator without bicarbonate, and glucose 2% *w/v*: RPMI-1640 2% G (Invitrogen, Madrid, Spain) was used for working cultures. pH was adjusted to 7.0 with 1 M sodium hydroxide and the resulting solution was filtered using a 0.22 µm filter. As samples were not sterile, chloramphenicol was added to the RPMI medium at a final concentration of 500 µg/mL. The yeast strain was first cultured on Sabouraud Dextrose agar (Invitrogen, Madrid, Spain) at 30 °C for 48 h before testing. The inoculums were prepared by suspending colonies in sterile distilled water to achieve a density equivalent to 2 McFarland standards, the counting of which took place in a Neubauer Chamber shortly afterwards (1 to 5 × $10^6$ Colony Forming Unit, CFU/mL). Working suspension was prepared by diluting the standardized suspension in sterile distilled water (1:10) in order to prepare 1 to 5 × $10^5$ CFU/mL. The test was performed using 96-well polystyrene sterile microdilution plates. Serial dilutions from 75 to 0.009 µg/mL of the following forms were compared. (1) Free-AmB (AmB solution first dissolved in DMSO at 300 µg/mL and afterwards in RPMI-1640 double strength at 150 µg/mL). (2) AmB-gel (AmB gel dissolved in RPMI-1640 double strength at 150 µg/mL). (3) Blank-gel (Blank-gel dissolved in RPMI-1640 double strength at 12.5–0.00075%).

Finally, 100 µL of inoculum was added to all the wells. In addition, 3 posts were reserved in each plate for both the positive control (300 µL of the inoculum) and negative control (300 µL of the culture broth). The plates were read at $t_0$ and then at 24 h and 48 h after incubation at 30 °C, with a microplate reader model 680 (Bio-Rad, Madrid, Spain) at $\lambda = 620$ nm.

### 2.10. *Atomic Force Microscopy (AFM)*

#### 2.10.1. Images

*C. albicans* cells were exposed to AMB-gel and Blank-gel during a period of 4 h. The cells were collected by centrifugation and washed three times with PBS solution (pH 7.40) in order to obtain a final suspension of 1 × $10^6$ yeast cells/mL. The suspension was filtered through SMWP01300 Millipore® filters (Merck Chemical and Life Science, S.A, Madrid, Spain) with a pore size similar to the yeast size followed by gentle rinsing with PBS in order to clean the filter from non trapped cells, then subsequently cut and attached to a steel disk using a small piece of double-sided adhesive tape. Finally, the sample was transferred to the AFM liquid cell while avoiding dewetting. Throughout the procedure, MSNL-10, V shaped silicon nitride cantilevers (Bruker AFM Probes, Camarillo, CA, USA) with a nominal spring constant of 0.03 N/m were used.

Prior to measuring the nanomechanics the spring constants of the cantilevers were determined by thermal noise method. AFM images were acquired in contact mode by minimizing the force during the scan and continuously adjusting the set point with a 0° scan angle at a scan rate of 1.5 Hz. Images were processed using NanoScope®® analysis software (Bruker, AXS Co., Madison, WI, USA).

#### 2.10.2. Force-Distance Curves

Mechanical properties were measured by recording arrays of 32 × 32 force curves, using a maximum force of 0.5–1 nN to avoid sample damage, a contact time of 100 ms, and approach and retract speeds of 1.0 µm/s. The Young's modulus of the yeast cells was determined as a first approximation by using the Hertz model.

### 2.11. *In Vivo Tolerance Study by Evaluating Biomechanical Properties of Human Skin*

Ten female volunteers between 25 and 35 years old with healthy skin participated in the study. The study was approved by the CEEA-UB according to the recommendations outlined in the Declaration of Helsinki (ref: IRB00003099) and all volunteers provided written informed consent [21]. All participants were requested not to use skin care cosmetics on the test areas of application for two days prior to the study. Skin temperature, transepidermal water loss (TEWL), and stratum corneum hydration (SCH) were determined using a thermometer ST500, a Tewameter TM 300, and a Corneometer CM 825 (Courage-Khazaka electronic GmbH, Cologne, Germany), respectively [22]. The measurements of these parameters were made before applying the AmB-gel (basal readings), immediately after the application of 0.5 mL/cm$^2$ ($t_0$) and after 2 h of application on the flexor side of the left forearm.

### 2.12. *In Vivo Tolerance Study by Draize Assay and Histological Analysis*

The tolerance of the formulation was evaluated using scarified rabbit skin. The blank-gel and AmB-gel were tested on New Zealand albino male rabbits (2 kg) according to the guidelines provided by the CEEA-UB. The rabbits were acclimated over a 5-day period before the study and were classified into three groups ($n$ = 3/group): Group A (Blank-gel), Group B (Skin scarified-control group), and Group C (AmB-gel). The surrounding area of the dorsal trunk was shaved with clippers where a square was drawn for scarification with a razor before beginning the assay. After 30 min, a volume of 0.5 mL of either Blank-gel or AmB-gel was topically applied on the scarified skin of each corresponding group while group B was not exposed to any treatment. This area was protected with gauze and secured with hypoallergenic sticking plaster for 48 h. The signal of edema and erythema were determined after 24 h and 48 h of exposure. Both scores were established according to the degree of severity and the primary irritation index value was calculated. The treatment was classified according the reported specifications: "nonirritant" (<0.5), "irritant" [2–5], or "highly irritant" [5–8,23]. Afterwards, the rabbits were anesthetized and euthanized with sodium pentobarbital.

For histological analysis, the samples of back skin from the rabbits were cut and set up for 24 h in 4% buffered formaldehyde at room temperature. After fixation, all samples were paraffin embedded in paraffin blocks, cut into 5 µm sections, and mounted on microscope slides. Afterwards, the samples were stained with hematoxylin and eosin and finally viewed on blind coded samples under a light microscope (Olympus BX41 and Olympus XC50 camera) with 100× magnification for the evaluation of the tissue structure.

### 2.13. Statistical Analysis

Obtained experimental data were analyzed by one-way analysis of variance (ANOVA). Comparisons of findings were done by multiple comparison test. A $p$ value $< 0.05$ was established as an indicator of statistically significant differences (SSD).

## 3. Results

### 3.1. Physicochemical Characterization

Figure 1, shows SEM micrographs of AmB-gel, which exhibited a heterogeneous structure with the formation of interconnected capillary channels in the form of holes similar to a porous sponge.

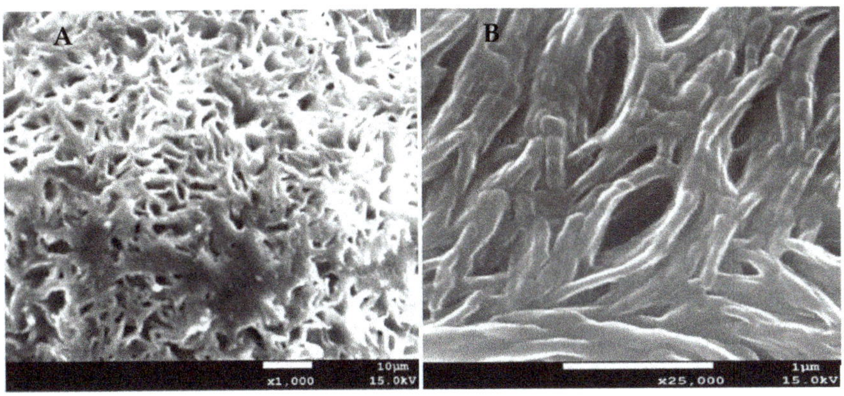


**Figure 1.** Morphological characterization by SEM. (**A**) AmB-gel porous sponge structure (×1000). (**B**) AmB-gel tubular appearance interconnected (×25,000).

The swelling process of AmB-gel followed a first-order kinetic model, which was represented by the kinetic constants k = 0.26 min$^{-1}$ ($r^2$ = 0.9986) (Figure 2). The degradation process of AmB-gel was completed in 20 min and followed a hyperbola model with a kinetic constant of 0.15 min$^{-1}$ ($r^2$ = 0.9975), whereas the percentage of P of AmB-gel was ~82.01 ± 0.5%.

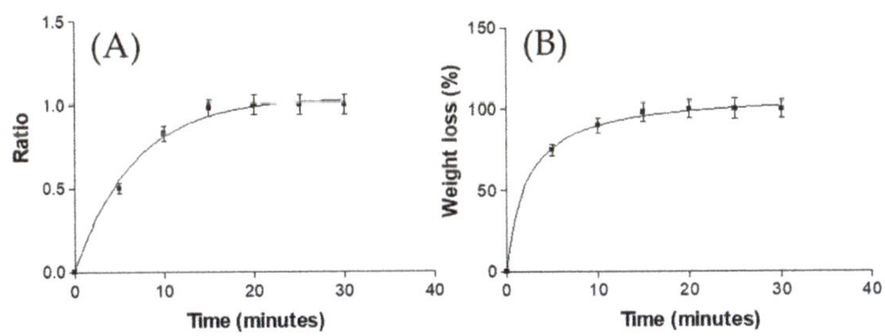


**Figure 2.** (**A**) Swelling ratio and (**B**) percentage of weight loss degradation of AmB-gel.

### 3.2. Stability Study

The pH value was suitable for application on the skin, ranging between 6.30 ± 0.05 and 6.70 ± 0.13 ($p > 0.05$) at 4 °C, and between 6.50 ± 0.13 and 6.67 ± 0.24 ($p > 0.05$) at 25 °C over a period of 90 days. The content of AmB present in the gel stored at 4 °C and 25 °C remained within the required margins of 90 to 100% during the first 90 days. With regards to the TurbiScan Lab®® analysis, Figure 3 shows the transmission profile of AmB-gel at 25 °C over a period of 90 days. The lower part of the vial is represented on the left side of the graph and the upper part on the right side. The peaks in the lower part and in the upper part of the vial correspond to the meniscus produced by the sample from contact with the glass. No sedimentation, flocculation, or coalescence phenomena were observed.

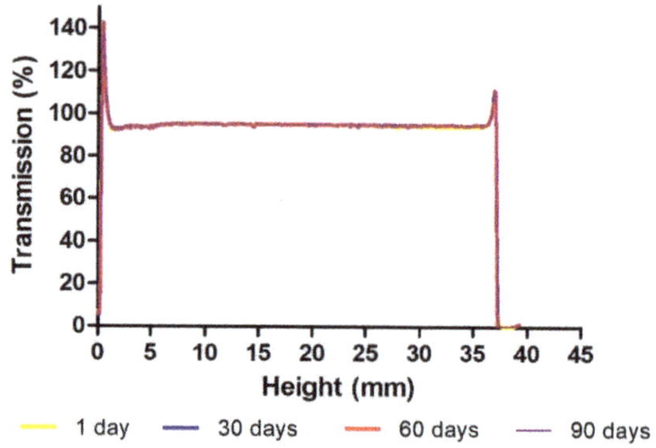


**Figure 3.** Transmission profile of AmB-gel obtained by Turbiscan Lab®® over 90 days at 25 °C. The left side of the curve corresponds to the bottom of the vial, whereas the right side corresponds to the sample behavior on the top.

## 3.3. *Rheological Analysis*

Figure 4A reproduces the flow curves and the viscosity curves obtained from the rheological characterization of AmB-gel. At 4 °C, the gel was liquid and showed a Newtonian behavior (constant viscosity under shear rate increase) adjusted by the Newton model ($r^2 = 1$) when under the programmed variations of shear rate. The viscosity at 4 °C was found to be 90.62 ± 7.112 × $10^{-2}$ mPa·s. Since AmB-gel transition is carried out as the temperature increases, at 32 °C a significantly sticky gel was formed which made it difficult to generate an adequate flow within the space between the conical plate. The rheogram (Figure 4B) showed a critical shear rate of ~10 $s^{-1}$. When above this value, the sample could not flow freely under the experimental conditions. As a result, the layout became quite irregular and the results were inconclusive. However, the valid curve portion showed non-Newtonian behavior with properties of shear thinning (pseudoplastic behavior), thus adjusting the Cross model ($r^2 = 0.981$). The punctual viscosity values at 10 $s^{-1}$ and 32 °C determined from the ramp up period was ~3.8 Pa·s. The estimated viscosity of the gel at 32 °C and 50 $s^{-1}$ was ~9.8 Pa·s.

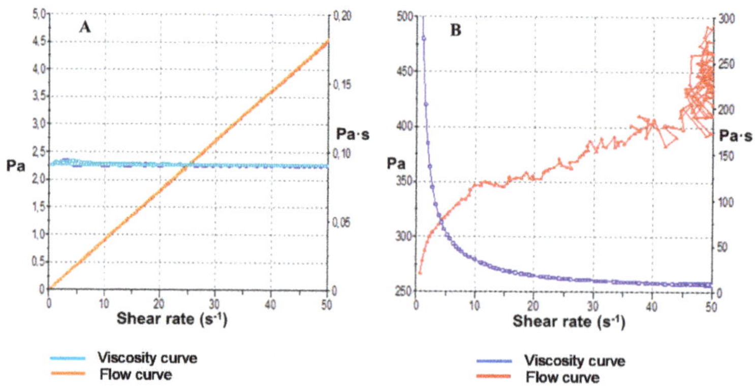


**Figure 4.** AmB-gel rheograms after 24 h: (**A**) Shear stress (Pa) and viscosity (Pa·s) curves at 4 °C and (**B**) Shear stress (Pa) and viscosity (Pa·s) curves at 32 °C.

### 3.4. Gelation Time and Spreadability

Gelation times of AmB-gel stored at 4 °C and 32 °C were 3 and 1.5 min, respectively. No differences in gelation times were evidenced between samples either with or without AmB. For AmB-gels stored for 3 months, the gelation times were even shorter at less than 1.5 min for 4 °C and 1 min 32 °C.

Figure 5 shows the results of spreadability test. The Boltzmann sigmoidal model for AmB-gel at 4 °C and hyperbola model for AmB-gel at 32 °C were the models that best fit the experimental data. At 4 °C, spreadability was greatly extended and with lower weight due to having lower viscosity at this temperature.

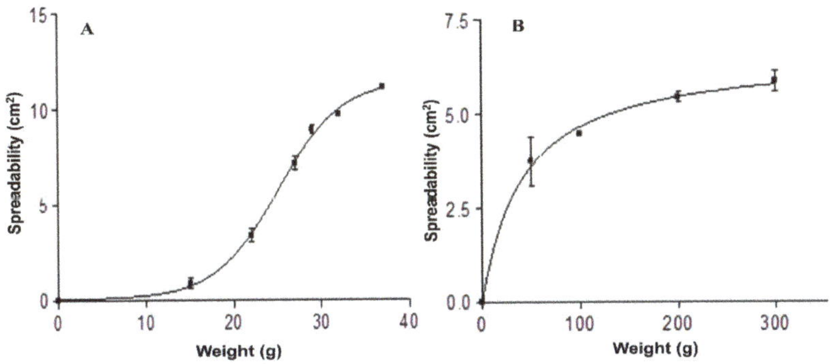


**Figure 5.** Spreadability charts of AmB-gel at different temperatures: (**A**) Spreadability curve at 4 °C; (**B**) Spreadability curve at 32 °C.

### 3.5. Release Studies

Figure 6 shows the release profile of AmB from the gel, where a sustained release exhibiting S-shaped behavior can be observed. After 24 h, more than 90% of the drug was released. The optimal kinetic model was fitted to the Boltzmann sigmoidal according to the following equation with AIC of 30.05 and $r^2$ of 0.999:

$$Y = Bottom + (Top - Bottom)/(1 + \exp\left(\frac{V50 - x}{Slope}\right) \quad (4)$$

where $Y$ is the amount of drug released and *Top* and *bottom* are the initial and final values of drug release. *V50* is the time it takes to release half of the maximum amount susceptible for release and the slope of the curve indicates the steepness.

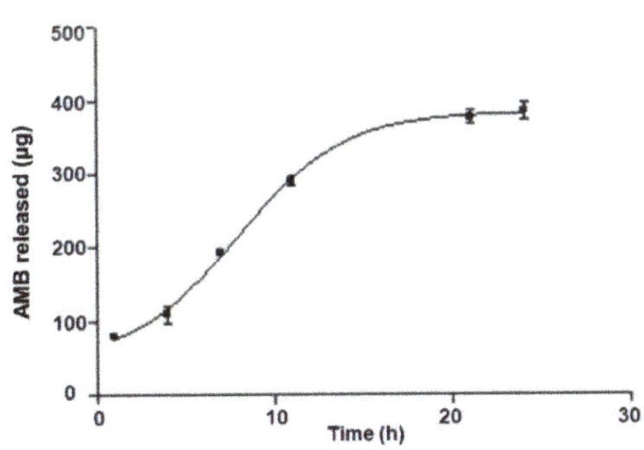


**Figure 6.** In vitro profile of AmB release from gel. The cumulative amount released was plotted against time. Data represent mean ± SD ($n$ = 6).

### 3.6. *Ex Vivo Permeation Studies of Skin and Vaginal Mucosa*

No amount of AmB from gel was detected in the receptor chamber in either skin or vaginal mucosa. Thus, neither flux nor permeation parameters could be calculated. The amount of AmB retained in the skin was 960.0 μg/g/cm$^2$ and vaginal mucosa 737.52 μg/g/cm$^2$.

### 3.7. *Antifungal Efficacy*

The values from the susceptibility test after 48 h are reported in Table 1 as MIC values against strains of *C. albicans*, *C. glabrata*, and *C. parapsilosis*. It can be observed that AmB-gel exhibited the lowest MIC values (0.09, 0.37 and 0.19 μg/mL) compared with Free-AmB.

**Table 1.** MIC against different cultures of *Candida* spp., Free-AmB, AmB-gel, and Blank-gel after incubation at 30 °C for 48 h ($n$ = 3).

| Tested Species | Origin | MIC (μg/mL) | | |
|---|---|---|---|---|
| | | Free-AmB | AmB-gel | Blank-gel |
| *C. albicans* | ATCC 10,231 | 0.15 | 0.09 | >250.000 |
| *C. glabrata* | ATCC 66,032 | 0.60 | 0.37 | >250.000 |
| *C. parapsilosis* | ATCC 22,019 | 0.30 | 0.19 | >250.000 |

### 3.8. *Atomic Force Microscopy*

Figure 7 shows the effect of a Blank-gel on *C. albicans* cell. Figure 7A shows the deflection image of a single yeast cell trapped in a pore which appears as the reddish region at the top-left of the image while the cell surface shows an extended region covered by filamentous structures. Figure 7B,C depicts magnified images of topographic and elasticity maps of this region prior to the addition of blank-gel to a final concentration of 6.5% (*w:v*) as well as Figure 7D,E after addition. The surface was quite smooth before the addition of the Blank-gel and some filamentous structures could be observed and differentiated by its different Young's modulus values with respect to the cell bulk structure (Figure 7B). After the addition of the blank-gel, the surface of the yeast cell became rougher and no filamentous structures could be observed any longer (Figure 7D). The elasticity map of the region (Figure 7E) evidences that it becomes more homogenous with the incorporation of polymer P407.

Figure 8 shows the effect of AmB-gel on *C. albicans* cells. Deflection, topographic and elasticity images (Figure 8A–C, respectively) of a single *C. albicans* yeast cell trapped in a pore are quite similar to those shown in Figure 6, although the filamentous structures are less evident in this case. The addition of AmB in the gel apparently altered the *C. albicans* cell. Figure 8D demonstrates that two different regions on cell surface can be clearly differentiated: (i) a reddish region (low region) that covers almost the 80% of the cell surface and (ii) a yellowish arched band (high region) crossing the surface from top-left to center-right of the cell surface protruding 40–80 nm over the reddish one. A close inspection of this yellowish band revealed a structure resembling a wrinkled sheet. Figure 8E,F shows the topographic and elasticity maps of the low region of Figure 8D. From these images, it is possible to recover the filamentous structures previously observed, although the differences in the Young's modulus values between the filaments and the bulk structure are not overly distinctive. Figure 8G,H shows the topographic and elasticity maps of the high region of Figure 8D. From the images, it is possible to observe two differentiated regions: (i) one diagonal from top-left corner to the bottom-right corner of Figure 8H (darker region) and (ii) a second region formed from the higher Young's values (lighter region) in Figure 8H occupying the bottom-left and top-right corners of the image with higher Young's modulus values than those of the low region. Considering that the darker region in Figure 8H is the higher region of the topographic image (Figure 8G), it is possible to identify these "wrinkled sheets" as regions with low elasticity modulus, thus suggesting that material in these structures is not as packed as those with blank-gel.

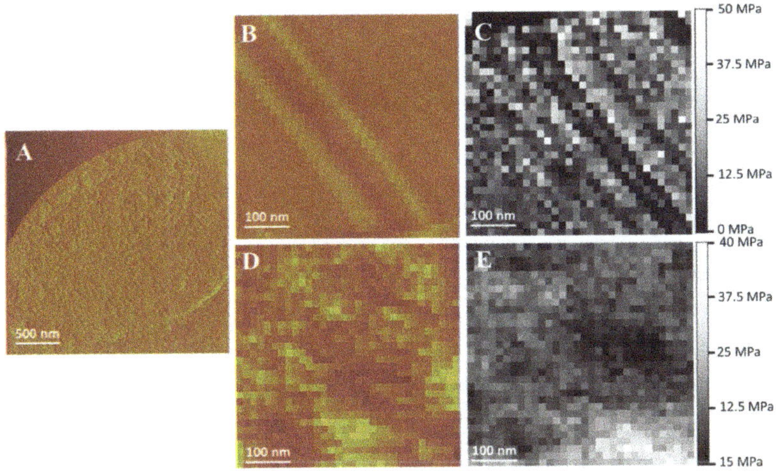


**Figure 7.** Effect of the Blank-gel on *C. albicans* cells. (**A**) Deflection image of a single yeast cell trapped in a pore. (**B**) Topographic map zoom image of the yeast cell before application of the Blank-gel. (**C**) Elasticity map zoom image of the yeast cell before application of the Blank-gel. (**D**) Topographic map zoom image of the cell yeast after application of the Blank-gel. (**E**) Elasticity map zoom image of the yeast cell after application of the blank-gel.

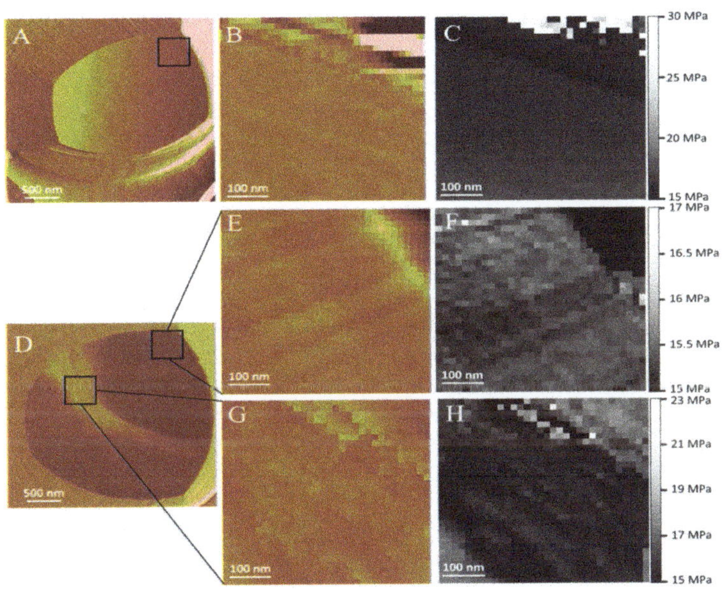


**Figure 8.** Effect of AmB-gel on *C. albicans* cells. (**A**) Deflection image of a single yeast cell trapped in a pore. (**B**) Topographic map zoom image of a single yeast cell trapped in a pore. (**C**) Elasticity map zoom image of a single yeast cell trapped in a pore. (**D**) A reddish region (low region) that covers almost the 80% of the cell surface and a yellowish arched band (high region) crossing the surface from top-left to center-right of the cell surface. (**E**) Topographic map zoom image of the low region. (**F**) Elasticity map zoom image of the low region. (**G**) Topographic map zoom image of the high region. (**H**) Elasticity map zoom image of the high region.

Conversely, Figure 9 shows the profiles of the elasticity values as a function of its frequency. Figure 9A represents the individual Young's modulus values as a function of its frequency before and after the addition of blank-gel. Before the addition of blank-gel, the yeast cell showed a wide distribution of Young's modulus values with a central value of 21.0 ± 0.4 MPa. After the addition of the blank-gel, the values distribution is narrower and centered at a higher value 26.29 × 0.08 MPa. Figure 9B shows a central peak around 18.48 MPa and a wide distribution function. The low region depicted in Figure 8D corresponds to a narrower peak with a Young's modulus mean value that is shifted towards smaller values (15.63 MPa). On the contrary, the protruding structure resembling "wrinkled sheets" in Figure 8D shows a bimodal distribution, one close to the values of the blank *C. albicans* cell with a mean Young's modulus value of 18.10 MPa and another shifted towards smaller values of 16.62 MPa.

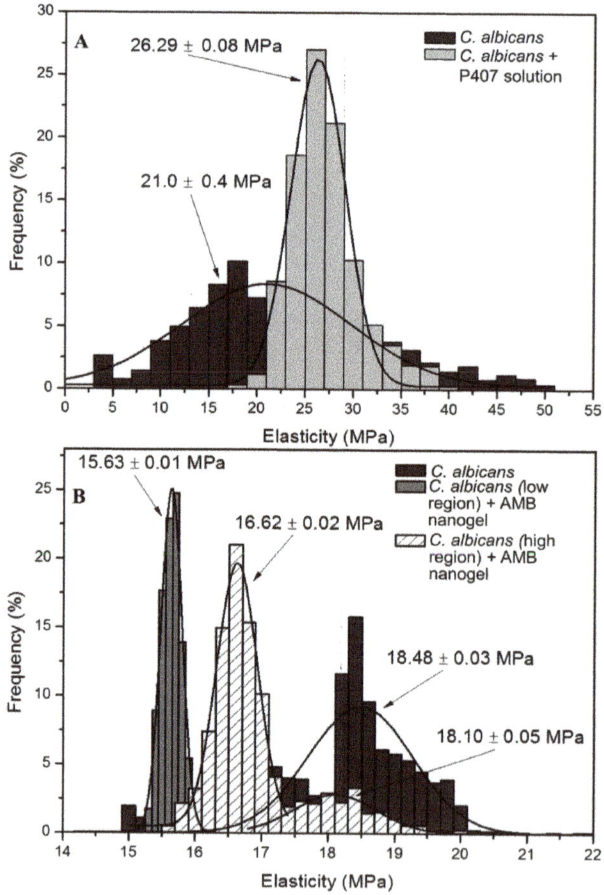


**Figure 9.** Individual Young's modulus values as a function of its frequency on *C. albicans* cells. (**A**) Individual Young's modulus values before and after the addition of Blank-gel. (**B**) Individual Young's modulus values before and after the addition of the AmB-gel.

### 3.9. *In Vivo Tolerance Study by Evaluating Biomechanical Properties of Human Skin*

The results of evaluated biomechanical parameters are depicted in Figure 10. No statistically significant changes of TEWL or skin temperature were observed between Blank-gel and AmB-gel or when compared with baseline measurements ($p > 0.05$). However, a significant decrease in SCH values was observed in the Blank-gel immediately after application (0 h) and at 2 h as well as in the AmB-gel after 2 h, with respect to the basal state (Figure 10E,F).

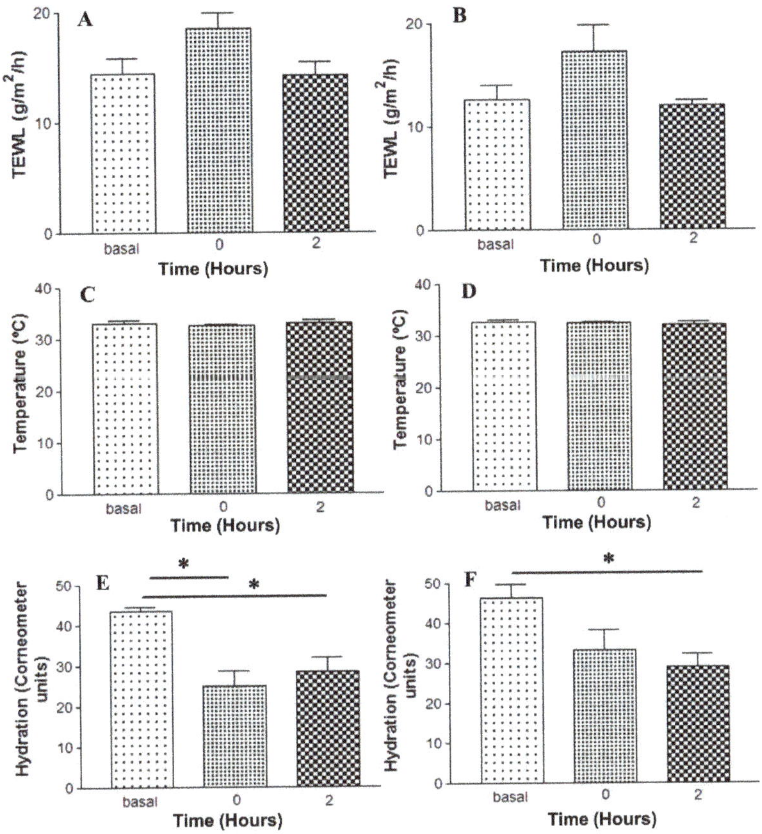


**Figure 10.** Biomechanical parameters evolution monitored before the application (basal), immediately after application (0 h) and 2 h after application. (**A**) TEWL values of blank-gel. (**B**) TEWL values of AmB-gel. (**C**) Skin temperature values of blank-gel. (**D**) Skin temperature values of AmB-gel. (**E**) SCH values of blank-gel. (**F**) SCH values of AmB-gel. * Statistically significant differences ($p < 0.05$).

### *3.10. In Vivo Tolerance Study on Scarified Rabbit Skin by Draize Assay and Histological Analysis*

The Draize test was carried out in order to evaluate the skin irritation potential of AmB-gel. After 48 h, the resulting primary irritation index value for Blank-gel and AmB-gel was 0.38 and 0.45, respectively. This result indicates that both blank-gel and AmB-gel are nonirritants.

Regarding the histological evaluation, micrographs revealed that scarification caused histological alterations and the presence of nonspecific inflammatory cells in the skin (Figure 11B) while the topical application of Blank-gel (Figure 11A) and AmB-gel (Figure 11C) notably repaired these alterations, resulting in a less pronounced inflammatory process than that of the control group (Figure 11B).

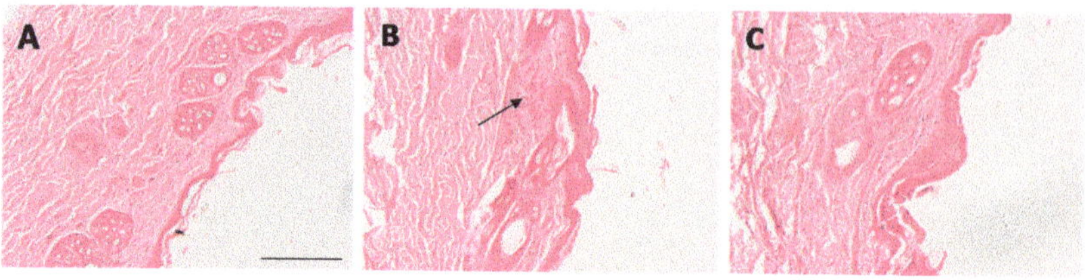


**Figure 11.** Optical microscopic images of skin. Blank-gel (**A**), skin-scarified control group (**B**), and AmB-gel (**C**). Hematoxylin and eosin stains nuclei blue/black while keratin and cytoplasm are stained red. Scale bar = 200 µm.

## 4. Discussion

P127 gels are used as the vanguard of drug delivery systems development due to the improved therapeutic efficacy and adherence to the recipient [24]. They have been used as vehicles for various routes of administration of drugs, including oral and topical [25], intranasal [18,26], vaginal and rectal [27], ocular [28], and parenteral [29]. In this study, AmB-gel was developed with 5% of DMSO as an optimal solubilizing agent and 25% of P127 to confer thermoreversible character. For the physicochemical characterization, SEM images were obtained, thus confirming that AmB-gel has a dense and well-oriented tubular appearance along with reticular networks. This type of structure has been reported in other studies [30,31], therefore supporting the postulation that a porous structure could help in the controlled release of drug.

The swelling of AmB-gel is a parameter dependent on the critical micelle concentration (CMC), given that P127-gels show two forms: gel form at 25 °C and liquid form at 4 °C, with liquid form showing thermoreversible behavior. Once the AmB-gel is poured into the swelling medium (PBS), a concentration of P407 decreases rapidly below CMC, thus producing swelling and solubilization in ~19 min. In the present study, 25% of P127 was used and at this concentration the SR followed a first-order kinetic model. Additionally, the degradation process was complete at 20 min, which is a typically rapid rate of degradation for these kinds of systems, as previous studies have also found [32]. On the other hand, the AmB-gel percentage of P was about 82.01 ± 0.5% corroborated with the SEM image which showed a high porosity structure.

The pH value of skin formulations is an important factor to consider in order to avoid skin irritation, particularly when under mycosis infections. AmB-gel showed pH levels to be slightly acidic, therefore biocompatibility with the natural acidity of the skin assures suitability for skin application [30]. What is more is that the AmB content in the gel was maintained between 90 and 100% over a span of 3 months while the superimposed graphs of transmission (Figure 3) showed variations below 10%, which indicates that the product will remain physically stable for a minimum of three months [33–35].

As expected, the rheological characteristics of AmB-gel were temperature-dependent (Figure 4). It is vital to evaluate rheological characterization because rheology can modulate biopharmaceutical properties such as release rates in addition to the widespread application on affected areas [36]. The viscosity at 4 °C was fairly low at ~90 mPa·s, which is an important value at the time of packaging because greater liquescence can lead to optimal forms of delivery, such as spray or roll-on applications, which help prevent the spread of mycotic infections. Once in contact with the affected skin at 32 °C, the solubility of the PPO chains decreases, subsequently forming micelles and assuming hexagonal and/or cubic structures, thus drastically increasing viscosity. The estimated viscosity of the AmB-gel at 32 °C and 50 $s^{-1}$ was ~9.8 Pa·s, which is nearly 100 times higher than the viscosity at 4 °C. This thermal gelation of P407 can form and sustain a drug depot in this area and subsequently increase the contact time, which can then produce desirable drug release characteristics and prolong pharmacological action [37]. The thermal gelation process did not seem to be affected by the presence of AMB. All the results suggested that gel state would be reached after application on the skin and also gain viscosity instantaneously to increase residence time of the formulation on the application area [13].

The data obtained from release studies demonstrated that the AmB-gel provided a sustained release for approximately 24 h. Model fitting showed that release mechanism followed a sigmoidal model. This faster release could be attributed to the highly porous structure and the rapid degradation of P407-gel. Moreover, this mathematical model has been associated with these types of gels due to the fact that they are used to simulate the transport and reaction of fluid in porous media [38].

Ex vivo permeation studies revealed that AmB did not permeate through human skin nor porcine vaginal mucosa. Similar results were reported in previous studies in which no AmB from a nanoemulsion was detected in the ex vivo permeation study using human skin or pig's ear skin [12,39], which might be explained by the great molecular weight (926 D) and high hydrophobicity of AmB, effectively limiting its ability to pass through the aqueous structure of the vaginal mucosa and the hydrophilic barrier of the dermis. However, it was observed that AmB had high retention in the skin and vaginal mucosa with values of 960 $\mu g/g/cm^2$ and 737 $\mu g/g/cm^2$, respectively. This result confirms that AmB succeeded in crossing the SC and was distributed to both the epidermis and dermis. Furthermore, it was able to permeate into the vaginal mucosa without reaching the systemic circulation, which was confirmed with the high amount of drug retained in the tissue and is indicative of the formulation favoring a local effect on the target area with no side effects. This high drug retention capacity in both tissues could be due to the presence of P407 in the formulation, since the chemical structure of this polymer allows it to act as a surfactant that enhances the diffusion ability of drug, resulting in increased drug concentrations in the tissue layer that leads to efficiently provide a local effect [18]. These results suggest that the AmB in this formulation could be successfully implemented in order to achieve a local effect on the skin or vaginal mucosa without adverse systemic effects.

The antifungal action of AmB-gel was clearly observed (Table 1) with a MIC value lower than Free-AmB in all tested Candida species and thus was highly effective—essentially confirming the adequacy of the formulation due to its improved effectiveness against fungus and yeast infections, which is likely due to the synergistic effects reported between the formulations of poloxamers and antifungal agents [40]. On the other hand, Blank-gel did not have any effect against Candida strains and produced values consistent with those reported in previous studies [12]. These results were subsequently observed by AFM images (Figures 7 and 8) which evidenced the alterations on the surface of Candida yeast cells that were induced by AmB-gel. In an equivocal manner, the mean values of Young's modulus (Figure 9) demonstrate the destructuring effect of AmB-gel on the outer membrane of the yeast cell.

The tolerance of the formulation was studied by evaluating the biomechanical properties of skin, since parameters such as TEWL and SCH are important indicators of skin integrity. Thus, any changes in these parameters could be directly related to an alteration of the functionality of the skin barrier [41]. From the results of this noninvasive in vivo method, no statistically significant differences were observed in temperature or TEWL values of volunteers after treatment with AmB-gel or with blank-gel when compared with baseline measurements (Figure 10), whereas SCH values showed a significant reduction in both AmB-gel and blank-gel with respect to the basal state. This decrease in the hydration levels in the area of application of the final product could be due to the ability of the gel to absorb moisture produced by the sweat glands. When the formulation containing P407 dries on the surface of the skin, it forms a film that captures moisture and does so in a state of relative equilibrium with ambient humidity. This effect would be especially suitable in skin candidiasis because it could prevent the proliferation of microorganisms around the affected area [42,43]. Moreover, when comparing the values of the biomechanical parameters evaluated for AmB-gel and blank-gel, the incorporation of the AmB does not significantly modify the properties of the gel. Furthermore, none of the volunteers exhibited side effects of burning or itching after the application. Therefore, the AmB-gel did not alter the biophysical properties of the skin of volunteers throughout the course of the study.

Finally, the tolerability of the formulation was evaluated by in vivo model using scarified rabbit skin in order to simulate the damage of the skin barrier caused by the fungal infection. The irritation index values for AmB-gel and Blank-gel were less than 0.5 and can therefore be classified as nonirritant formulations. This result was consistent with histological analysis, which showed that both formulations did not cause any alteration of the skin architecture but, on the contrary, they actually repaired the ulcerated skin. According to this finding the ability of P407 to reduce wounds in burn patients while simultaneously stimulating the proliferation of collagen fibers, consequently increasing scarring and generating new tissue, have all been reported in previous studies [44]. In conjunction, these effects might favor the healing and alleviation of wounds caused by Candida. In conclusion, AmB-gel represents a promising therapeutic option to treat these types of infections via dermal application.

## 5. Conclusions

The focus of the present study was on the incorporation of AmB in a copolymer P407-based gel designed for application on both the skin and vaginal mucosa, specifically for the treatment of candidiasis. P407 has shown great potential as a support system in the administration of hydrophilic and hydrophobic drugs, as is the case of AmB. P407 is suitable due to its thermoreversible capacity, low toxicity, and permeation-enhancing properties. During the experimental phase, AmB-gel remained stable for at least 3 months, thereby demonstrating pH values suitable for both skin and vaginal application. Due to its low viscosity at predetermined temperatures and its porous structure which facilitates rapid drug release, AmB-gel can also be used in aerosol form in order to avoid contact with areas infected by the fungus and thus prevent the spread of infection to other anatomical sites while minimizing the risk of a subsequent superinfection. No drug was quantified in the receptor compartments of the Franz cells, which indicates nonabsorption of AmB and therefore avoids adverse systemic effects. None of volunteers exhibited burning or itching after application and the histological images of rabbit skin displayed high tolerability of the formulation. Furthermore, AmB-gel did not alter the biophysical properties of skin, thus confirming adequate safety as topical agents. According to the values of the amount of drug retained in the skin and vaginal mucosa, along with the MIC values and the images obtained by AFM, it can be concluded that AmB-gel provides a satisfactory antifungal effect across various Candida species.

**Author Contributions:** L.S. carried out the experiments and analyzed and wrote the results of the manuscript. A.C.C. conceived the idea and analyzed the results. M.S.-A. analyzed the results and wrote and edited the manuscript. M.R. and L.C.E. participated in the methodologic analysis and editing of the manuscript. M.J.R.-L. realized the histological studies. N.B. carried out the microbiological efficacy. O.D. performed the force atomic experiments and analyzed the results. B.C. conceived the idea and edited the manuscript. All authors read the manuscript and agreed with the submission.

**Funding:** This research was funded by L'Agència de Gestiód'Ajuts Universitarisi de Recerca (AGAUR) grant number [SGR-2017 1744]. Lilian Sosa also expresses gratitude for the grant awarded by the Institute of Nanoscience and Nanotechnology IN2UB number [2017.3.IN2UB.2] in the completion of her doctoral thesis. Marcelle Silva-Abreu acknowledges the support from the Coordination for the Improvement Personnel (CAPES-Brazil).

**Acknowledgments:** We appreciate the help of Jonathan Proctor for the revision of the use of the English language. The authors are grateful to Lyda Halbaut for the help with rheological studies.

**Conflicts of Interest:** The authors declare no conflicts of interest.

## References

1. Havlickova, B.; Czaika, V.A.; Friedrich, M. Epidemiological trends in skin mycoses worldwide. *Mycoses* **2008**, *51*, 2–15. [CrossRef] [PubMed]
2. Sawant, B.; Khan, T. Recent advances in delivery of antifungal agents for therapeutic management of candidiasis. *Biomed. Pharm.* **2017**, *96*, 1478–1490. [CrossRef] [PubMed]
3. Sanchez, D.A.; Schairer, D.; Tuckman-Vernon, C.; Chouake, J.; Kutner, A.; Makdisi, J.; Friedman, J.M.; Nosanchuk, J.D.; Friedman, A.J. Amphotericin B releasing nanoparticle topical treatment of Candida spp. in the setting of a burn wound. *Nanomedicine* **2014**, *10*, 269–277. [CrossRef] [PubMed]

4. Qasim, M.; Baipaywad, P.; Udomluck, N.; Na, D.; Park, H. Enhanced therapeutic efficacy of lipophilic amphotericin B against Candida albicanswith amphiphilic poly(N-isopropylacrylamide) nanogels. *Macromol. Res.* **2014**, *22*, 1125–1131. [CrossRef]
5. Yapar, N. Epidemiology and risk factors for invasive candidiasis. *Clin. Risk Manag.* **2014**, *10*, 95–105. [CrossRef]
6. Cannon, R.D.; Lamping, E.; Holmes, A.R.; Niimi, K.; Baret, P.V.; Keniya, M.V.; Tanabe, K.; Niimi, M.; Goffeau, A.; Monk, B.C. Efflux-mediated antifungal drug resistance. *Clin. Microbiol. Rev.* **2009**, *22*, 291–321. [CrossRef] [PubMed]
7. Canuto, M.M.; Rodero, F.G. Antifungal drug resistance to azoles and polyenes. *Lancet Infect. Dis.* **2002**, *2*, 550–563. [CrossRef]
8. Morio, F.; Jensen, R.H.; Le Pape, P.; Arendrup, M.C. Molecular basis of antifungal drug resistance in yeasts. *Int. J. Antimicrob. Agents* **2017**, *50*, 599–606. [CrossRef]
9. Miceli, M.H.; Díaz, J.A.; Lee, S.A. Emerging opportunistic yeast infections. *Lancet Infect Dis.* **2011**, *11*, 142–151. [CrossRef]
10. Cannon, R.D.; Lamping, E.; Holmes, A.R.; Niimi, K.; Tanabe, K.; Niimi, M.; Monk, B.C. Candida albicans drug resistance another way to cope with stress. *Microbiology* **2007**, *153*, 3211–3217. [CrossRef]
11. Vincent, B.M.; Lancaster, A.K.; Scherz-Shouval, R.; Whitesell, L.; Lindquist, S. Fitness trade-offs restrict the evolution of resistance to amphotericin B. *PLoS Biol.* **2013**, *11*, e1001692. [CrossRef] [PubMed]
12. Sosa, L.; Clares, B.; Alvarado, H.L.; Bozal, N.; Domenech, O.; Calpena, A.C. Amphotericin B releasing topical nanoemulsion for the treatment of candidiasis and aspergillosis. *Nanomedicine* **2017**, *13*, 2303–2312. [CrossRef] [PubMed]
13. Tutaj, K.; Szlazak, R.; Szalapata, K.; Starzyk, J.; Luchowski, R.; Grudzinski, W.; Osinska-Jaroszuk, M.; Jarosz-Wilkolazka, A.; Szuster-Ciesielska, A.; Gruszecki, W.I. Amphotericin B-silver hybrid nanoparticles: Synthesis, properties and antifungal activity. *Nanomedicine* **2016**, *12*, 1095–1103. [CrossRef] [PubMed]
14. Perez, A.P.; Altube, M.J.; Schilrreff, P.; Apezteguia, G.; Celes, F.S.; Zacchino, S.; de Oliveira, C.I.; Romero, E.L.; Morilla, M.J. Topical amphotericin B in ultradeformable liposomes: Formulation, skin penetration study, antifungal and antileishmanial activity in vitro. *Colloids Surf. B Biointerfaces* **2016**, *139*, 190–198. [CrossRef] [PubMed]
15. Atiyeh, B.S.; Gunn, S.W.; Hayek, S.N. State of the art in burn treatment. *World J. Surg.* **2005**, *29*, 131–148. [CrossRef] [PubMed]
16. Giuliano, E.; Paolino, D.; Fresta, M.; Cosco, D. Mucosal application of polaxamer 407-based hydrogels: An overview. *Pharmaceutics* **2018**, *10*, 159. [CrossRef] [PubMed]
17. Dumortier, G.; Grossiord, J.L.; Agnely, F.; Chaumeil, J.C. A review of poloxamer 407 pharmaceutical and pharmacological characteristics. *Pharm. Res.* **2006**, *23*, 2709–2728. [CrossRef]
18. Espinoza, L.C.; Silva-Abreu, M.; Clares, B.; Rodríguez-Lagunas, M.J.; Halbaut, L.; Cañas, M.A.; Calpena, A.C. Formulation Strategies to Improve Nose-to-Brain Delivery of Donepezil. *Pharmaceutics* **2019**, *11*, 64. [CrossRef]
19. Subcommittee on Antifungal Susceptibility Testing (AFST) of the ESCMID European Committee for Antimicrobial Susceptibility Testing (EUCAST). EUCAST Definitive Document EDef 7.1: Method for the determination of broth dilution MICs of antifungal agents for fermentative yeasts. *Clin. Microbiol. Infect.* **2008**, *14*, 398–405. [CrossRef]
20. Clinical and Laboratory Standards Institute, CLSI. *Reference Method for Broth Dilution Antifungal Susceptibility Testing of Yeast*, 3rd ed.; document M27-A3; Clinical and Laboratory Standards Institute: Villanova, PA, USA, 2008; Volume 28, ISBN 1-56238-666-2, ISSN 0273-3099.
21. World Medical Association Declaration of Helsinki. Clinical Review and Education. 2013. Available online: https://www.wma.net/wp-content/uploads/2016/11/DoH-Oct2013-JAMA.pdf (accessed on 2 May 2019).
22. Du Plessis, J.; Stefaniak, A.; Eloff, F.; John, S.; Agner, T.; Chou, T.C.; Nixon, R.; Steiner, M.; Franken, A.; Kudla, I.; et al. International guidelines for the in vivo assessment of skin properties in non-clinical settings: Part 2. transepidermal water loss and skin hydration. *Ski. Res. Technol.* **2013**, *19*, 265–278. [CrossRef]
23. Draize, J.H.; Woodward, G.; Calvery, H.O. Methods for the study of irritation and toxicity of substances applied topically to the skin and mucous membranes. *J. Pharmacol. Exp. Ther.* **1944**, *82*, 377–390.
24. Wertheimer, A.I.; Santella, T.M.; Finestone, A.J.; Levy, R.A. Drug delivery systems improve pharmaceutical profile and facilitate medication adherence. *Adv. Ther.* **2005**, *22*, 559–577. [CrossRef] [PubMed]
25. Padilla, M.; Clark, G.T.; Merrill, R.L. Topical medications for orofacial neuropathic pain: A review. *J. Am. Dent. Assoc.* **2000**, *131*, 2. [CrossRef]

26. Jain, N.K.; Shah, B.K.; Taneja, L.N. Nasal absorption of metoprolol tartrate. *Indian J. Pharm. Sci.* **1991**, *53*, 16–19.
27. Ryu, J.M.; Chung, S.-J.; Lee, M.-H.; Chang-Kook, K.; Chang-Koo, S. Increased bioavailability of propanolol in rats by retaining thermally gelling liquid suppositories in the rectum. *J. Control. Release* **1999**, *59*, 163–172. [CrossRef]
28. Li, H.; Sung, K.C. Carbopol/pluronic phase change solutions for ophthalmic drug delivery. *J. Control. Release* **2000**, *69*, 379–388.
29. Koller, C.; Buri, P. Propriétés et intérêtpharmaceutique des gels thermoréversibles à base de poloxamers et poloxamines. *STP Pharma* **1987**, *3*, 115–124.
30. Yeh, M.Y.; Zhao, J.Y.; Hsieh, Y.R.; Lin, J.H.; Chen, F.Y.; Chakravarthy, R.D.; Chung, P.C.; Lin, H.C.; Hung, S.C. Reverse thermo-responsive hydrogels prepared from Pluronic F127 and gelatin composite materials. *RSC Adv.* **2017**, *7*, 21252. [CrossRef]
31. Rangabhatla, A.S.L.; Tantishaiyakul, V.; Boonrat, O.; Hirun, N.; Ouiyangkul, P. Novel in situ mucoadhesive gels based on Pluronic F127 and xyloglucan containing metronidazole for treatment of periodontal disease. *Iran. Polym. J.* **2017**, *26*, 851–859. [CrossRef]
32. Mallandrich, M.; Fernández-Campos, F.; Clares, B.; Halbaut, L.; Alonso, C.; Coderch, L.; Garduño-Ramírez, M.L.; Andrade, B.; Del Pozo, A.; Lane, M.E.; et al. Developing transdermal Application of Ketorolac Tromethamine Entrapped in Stimuli Sensitive Block Copolymer Hydrogels. *Pharm. Res.* **2017**, *34*, 1728–1740. [CrossRef]
33. Sierra, A.F.; Ramirez, M.L.; Campmany, A.C.; Martinez, A.R.; Naveros, B.C. In vitro and In vivo evaluation of the use of a newly developed melatonin loaded emulsion combined with UV filtres as a protective agent against skin irradiation. *J. Derm. Sci.* **2013**, *69*, 202–214. [CrossRef] [PubMed]
34. Alvarado, H.L.; Abrego, G.; Souto, E.B.; Garduño-Ramírez, M.L.; Clares, B.; García, M.L.; Calpena, A.C. Nanoemulsions for dermal controlled release of oleanolic and ursolic acids: In vitro, ex vivo and in vivo characterization. *Colloids Surf. B Biointerfaces* **2015**, *130*, 40–47. [CrossRef] [PubMed]
35. Abrego, G.; Alvarado, H.L.; Egea, M.A.; Gonzalez-Mira, E.; Calpena, A.C.; Garcia, M.L. Design of nanosuspensions and freeze-dried PLGA nanoparticles as a novel approach for ophthalmic delivery of pranoprofen. *J. Pharm. Sci.* **2014**, *103*, 3153–3164. [CrossRef] [PubMed]
36. Fernandez-Campos, F.; Clares Naveros, B.; Lopez Serrano, O.; Alonso Merino, C.; Calpena Campmany, A.C. Evaluation of novel nystatin nanoemulsion for skin candidosis infections. *Mycoses* **2013**, *56*, 70–81. [CrossRef] [PubMed]
37. Lenaerts, V.; Triqueneaux, C.; Quartern, M.; Rieg-Falson, F.; Couvreur, P. Temperature-dependent rheological behavior of Pluronic F-127. *Int. J. Pharm.* **1987**, *39*, 121–127. [CrossRef]
38. Navarro-Verdugo, A.L.; Goycoolea, F.M.; Romero-Melendez, G.; Higuera-Ciaparad, I.; Arguelles-Monal, W. A modified Boltzmann sigmoidal model for the phase transition of smart gels. *Soft Matter* **2011**, *7*, 5847. [CrossRef]
39. Moraes Santos, C.; Barbosa de Oliveira, R.; Arantes, V.T.; Rodrigues-Caldeira, L.; De Oliveira, M.C.; Tabosa-Egito, E.S.; Miranda-Ferreira, L.A. Amphotericin B-Loaded Nanocarriers for Topical Treatment of Cutaneous Leishmaniasis: Development, Characterization, and In Vitro Skin Permeation Studies. *J. Biomed. Nanotechnol.* **2012**, *8*, 322–329.
40. Banov, D. Synergistic Effect of Poloxamer-based Composition and Itraconazole on Fungus and Yeast. U.S. Patent 9,763,934B2, 19 September 2017.
41. Mohammed, D.; Hirata, K.; Hadgraft, J.; Lane, M.E. Influence of skin penetration enhancers on skin barrier function and skin protease activity. *Eur. J. Pharm. Sci.* **2014**, *51*, 118–122. [CrossRef]
42. Gilbert, J.; Richardson, J.; Pallin, K.J.; Hadgraft, J.T. The effect of solutes and polymers on the gelation properties of Pluronic F-127 solution for controlled drug delivery. *J. Control. Release* **1987**, *5*, 113–118. [CrossRef]
43. Grove, G.L.; Allenby, K.; Houser, T. DFD-01 Reduces Transepidermal Water Loss and improves skin hydration and flexibility. *Dermatol. Ther.* **2017**, *7*, 507–514.
44. Kant, V.; Gopal, A.; Kumar, D.; Gopalkrishnana, A.; Pathaka, N.N.; Kurade, N.P.; Tandana, S.K.; Dinesh, K. Topical pluronic F-127 gel application enhances cutaneous wound healing in rats. *Acta Histochem.* **2013**, *116*, 5–13. [CrossRef] [PubMed]

*pharmaceutics*

MDPI

Article

# Repurposing Butenafine as An Oral Nanomedicine for Visceral Leishmaniasis

**Adriana Bezerra-Souza [1], Raquel Fernandez-Garcia [2], Gabriela F. Rodrigues [1], Francisco Bolas-Fernandez [3], Marcia Dalastra Laurenti [1], Luiz Felipe Passero [1,4], Aikaterini Lalatsa [5] and Dolores R. Serrano [2,*]**

1 Laboratory of Pathology of Infectious Diseases (LIM-50), Medical School, University of São Paulo, Avenida Dr. Arnaldo 455, 01246903 Cerqueira César, SP, Brazil

2 Departament of Pharmaceutics and Food Technology and Instituto Universitario de Farmacia Industrial (IUFI), School of Pharmacy, University Complutense, Avenida Complutense, 28040 Madrid, Spain

3 Departament of Microbiology and Parasitology, School of Pharmacy, Universidad Complutense de Madrid, Plaza Ramón y Cajal s/n, 28040 Madrid, Spain

4 Institute of Biosciences, São Paulo State University (UNESP), São Vicente. Praça Infante Dom Henrique, s/n, 11330-900 São Vicente, SP, Brazil

5 Institute of Biomedical and Biomolecular Sciences, School of Pharmacy and Biomedical Sciences, University of Portsmouth, White Swan Road, Portsmouth PO1 2 DT, UK

* Correspondence: drserran@ucm.es; Tel.: +34-91-394-1620

Received: 30 June 2019; Accepted: 18 July 2019; Published: 20 July 2019

**Abstract:** Leishmaniasis is a neglected tropical disease affecting more than 12 million people worldwide, which in its visceral clinical form (VL) is characterised by the accumulation of parasites in the liver and spleen, and can lead to death if not treated. Available treatments are not well tolerated due to severe adverse effects, need for parenteral administration and patient hospitalisation, and long duration of expensive treatments. These treatment realities justify the search for new effective drugs, repurposing existing licensed drugs towards safer and non-invasive cost-effective medicines for VL. In this work, we provide proof of concept studies of butenafine and butenafine self-nanoemulsifying drug delivery systems (B-SNEDDS) against *Leishmania infantum*. Liquid B-SNEDDS were optimised using design of experiments, and then were spray-dried onto porous colloidal silica carriers to produce solid-B-SNEDDS with enhanced flow properties and drug stability. Optimal liquid B-SNEDDS consisted of Butenafine:Capryol 90:Peceol:Labrasol (3:49.5:24.2:23.3 *w/w*), which were then sprayed-dried with Aerosil 200 with a final 1:2 (Aerosil:liquid B-SNEDDS *w/w*) ratio. Spray-dried particles exhibited near-maximal drug loading, while maintaining excellent powder flow properties (angle of repose <10°) and sustained release in acidic gastrointestinal media. Solid-B-SNEDDS demonstrated greater selectivity index against promastigotes and *L. infantum*-infected amastigotes than butenafine alone. Developed oral solid nanomedicines enable the non-invasive and safe administration of butenafine as a cost-effective and readily scalable repurposed medicine for VL.

**Keywords:** butenafine; SNEDDS; solid SNEDDS; spray drying; leishmaniasis; design of experiments

## 1. Introduction

Leishmaniasis is an infectious disease caused by parasites belonging to the *Leishmania* genus. The prevalence of leishmaniasis exceeds 12 million cases, and it is endemic in 98 countries in five continents. *Leishmania* parasites are transmitted by insect vectors from the genus *Lutzomyia* sp. or *Psychodopygus* sp. in the New World and *Phlebotomus* sp. in the Old World [1,2]. Leishmaniasis presents in the cutaneous (CL) and visceral (VL) leishmaniasis forms, depending on the type of host immune response and infecting parasite species [3,4]. In the New World, parasites of the subgenus *Vianna* cause only CL and mucocutanoues leishmaniasis (MCL) while parasites of the subgenus *Leishmania* are responsible for CL and VL [5]. VL is a chronic disease caused by *L. (L.) infantum* and *L. (L.) donovani* species [6–8] residing in host macrophages, mainly from spleen, liver, bone marrow, and lymph nodes, and is characterized by prolonged fever, hepatosplenomegaly, lymphadenopathy, anemia with leukopenia, hypergammaglobulinemia and hypoalbuminemia, weight loss, edema, and a debilitating state leading to weakening and ultimately death if untreated [6].

VL treatment mainstays involve pentavalent antimonials (SbV) and amphotericin B (AmB) as the first line in the developing and developed world, respectively [9]. Although they are both highly efficacious in vivo, SbV are linked to severe and frequent side effects, limiting their use [9], combined with a high rate of clinical resistance [10]. AmB, on the other hand, has shown limited resistance, but its clinical use is limited by the high cost, especially for safer lipidic formulations, such as Ambisome®, thermal instability, and nephrotoxicity [11,12]. Miltefosine, the only oral VL licensed treatment (licensed in 2003 in India), has several limitations as monotherapy, as its activity is highly dependent on the clinical form of leishmaniasis and the parasite strain, and patients frequently experience severe gastrointestinal disorders. In addition, it is teratogenic, and thus cannot be used in women of child-bearing age [13,14]. Miltefosine has a long elimination half-life (seven days) and a narrow therapeutic index, characteristics that limit the administered dose, which can lead to subtherapeutic levels over several weeks, encouraging the emergence of resistance [12,14,15]. Indeed, resistance has been reported in India (a country that alone accounts for 50% of the VL worldwide burden) and France [12]. Thus, available medicines for VL are outdated, impractical, insufficiently efficacious, or subject to resistance and unacceptable toxicities. Ideal treatments for VL should be able to possess greater than 95% efficacy, be orally administered, be stable in a tropical environment, be affordable, and require fewer than 11 days of treatment with an excellent side-effect profile [12,16].

Repurposing drugs is a strategy that has been crowned with success, as repurposed drugs have made up a third of all new commercially and clinically used drug treatments since 2009 [17,18]. Current VL drugs are interesting examples of repurposed drugs, since SbV, AmB, and miltefosine were originally used as an emetic, antifungal, and anticancer drug, respectively [14,19]. Butenafine is an allylamine drug, commonly employed in the treatment of fungal skin infections, such as ringworm, athlete's foot, jock itch, and pityriasis. Currently, it is only commercialised as a 1% cream for topical use. We have shown recent reports on how butenafine repositioning was effective in vitro against *L. (L.) amazonensis* and *L. (V.) braziliensis*, ethiological agents of CL and MCL. Furthermore, parasite treated with butenafine showed morphological alterations that resembled programmed cell death, which is attributed to the blockage of the biosynthesis of ergosterol [20]. Butenafine has limited oral bioavailability, with 1.5–3% of the oral dose being recovered in the plasma an hour after a single oral dosing of radiolabeled butenafine (0.2 mg/kg) in rats [21]. Butenafine is highly metabolized in the liver (methylation, dealkylation, and hydroxylation) and only 0.03% of the oral dose has been recovered intact from the plasma after 4 h [21]. This concurs with levels of its major metabolite (1-napthoic acid) in the plasma, which ranged between 1%–100% of administered parent drug dose [21].

Here, we are reporting the development of an oral butenafine nanomedicine able to enhance its aqueous solubility, maintaining a solubilized state in the gastrointestinal to allow for enhanced oral absorption, in order to target the liver and spleen (i.e., organs where the *Leishmania* parasite resides in high concentration). We have shown that SNEDDS are able to enhance the oral bioavailability of poorly soluble drugs and enable therapeutic concentrations to be delivered in the liver and spleen [12]. Thus, we hypothesised that if butenafine is formulated with GRAS (Generally Regarded as Safe) excipients with known activity against different *Leishmania* strains [22], we can develop butenafine-loaded SNEDDS (B-SNEDDS) and solid SNEDDS (solid B-SNEDDS) with enhanced activity, as well as being able to maintain butenafine's oral solubilisation capacity in the gastrointestinal tract. To ensure that a stable, and ideally solid, cost-effective formulation is available, we used design of experiments (DoE) studies to prepare butenafine SNEDDS colloidal silicon dioxide spray-dried particles that can be easily compressed into cost-effective, easily scalable, solid dosage forms of a repurposed drug for VL.

## 2. Materials and Methods

### 2.1. Materials

Butenafine hydrochloride (purity ≥ 98%) was purchased from Cayman Chemical Co. (Michigan, MI, USA). SNEDDS excipients (Capryol 90 (propylene glycol monocaprylate), Labrafil M 1944 CS (oleoyl polyoxyl-6 glycerides), Labrasol (caprylocaproyl polyoxyl-8 glycerides), and Peceol (glyceryl monooleate)) were kindly donated by Gattefosse (Saint-Priest Cedex, France). Two Aerosil silicon dioxide excipients were used as inert solid carriers: Aerosil R972 from Degussa (Frankfurt, Germany) and Aerosil 200 from Evonik Industries (Darmstadt, Germany). All other chemicals were of ACS reagent grade (Sigma Aldrich, Madrid, Spain) and were used as supplied. Solvents were of HPLC grade (Fisher, Madrid, Spain).

### 2.2. Solubility Studies of Butenafine

An excess quantity of butenafine was added to each of the excipients used in the preparation of SNEDDS, and the mixture was shaken (300 rpm) overnight at 25 °C in triplicate. The mixtures were centrifuged at 3000 rpm for 5 min, and the supernatant (0.1 mL) was diluted with 10 mL of methanol. The absorbance was measured in a spectrophotometer (JASCO V-730 spectrophotometer Madrid, Spain) at 220 nm to determine the solubility of butenafine. A calibration curve was performed previously in methanol to establish the linearity between concentration and absorbance at 220 nm.

### 2.3. Pseudo-Ternary Phase Diagrams

Ternary phase diagrams were constructed to study the phase behaviour of oils/surfactants over the whole concentration range. The existence of a microemulsion (type II) region within this diagram was observed visually. A D-Optimal design was developed using different mixtures of Capryol 90, Labrasol, and Peceol by using Design Expert software (State Ease, Minneapolis, MN, USA). Mixtures were vortexed for 5 min. The particle size distribution (PSD) was measured after dilution in deionised water (1:1000 *v/v*) in a Microtrac Zetatrac (Microtrac, Montgomeryville, PA, USA). The optimal excipient combination leading to the smallest PSD upon dilution was selected, in order to solubilise the drug and perform further experiments.

### 2.4. Preparation of Liquid B-SNEDDS Formulations

Based on solubility and phase diagram studies, Labrasol, Capryol 90, and Peceol were selected as a high-hydrophilic-lipophilic balance (HLB) surfactant, a medium-HLB surfactant, and an oil, respectively. Optimal SNEDDS were prepared combining the three as Capryol 90:Labrasol:Peceol (51:24:25 *w/w*). Butenafine (30 mg/g) was solubilised in the resulting excipient mixture, which was stirred overnight in order to obtain a homogenous isotropic mixture.

### 2.5. Preparation of Solid B-SNEDDS Formulation

A mini-spray dryer (Büchi B-191) was employed for the preparation of solid B-SNEDDS. Aerosil 200 (hydrophilic fumed amorphous silica, 5–50 nm (Evonik industries, Darmstadt, Germany)) or Aerosil R972 (hydrophobic fumed amorphous silica, 16 nm (Degussa AG, Frankfurt, Germany)) were used as inert carriers. A carrier (1 g) was dispersed in 100 mL of ethanol by magnetic stirring, after which liquid B-SNEDDS formulation (0.5 g) was mixed. The resulting suspension was delivered to a two-fluid nozzle (0.7 mm nozzle tip and a 1.5 mm diameter nozzle screw cap) using a peristaltic pump, at a speed of 10% (5 mL/min). Compressed air (2 bars) was used as the drying gas in a co-current mode, with the aspirator capacity set to maximum (100%). The flow-meter for the standard two-fluid nozzle was set to 600 NL/h. The inlet temperature was fixed at 62 °C, and the outlet temperature varied between 32 and 36 °C. After spray-drying, the dry powder was collected from the collector vessel, and the yield of the process was quantified using the following equation:

$$Yield\ (\%) = \frac{Collected\ mass\ after\ spray\ drying\ (mg)}{Total\ mass\ spray\ dried\ (mg)} \times 100 \quad (1)$$

A $2^2$ DoE was performed to understand and optimise key formulation parameters affecting the preparation of solid B-SNEDDS. The surface properties of silica used, i.e., type of Aerosil (200 or R972), as well as the weight ratio between Aerosil and liquid B-SNEDD (1:2 or 1:3 *w*/*w*) on the physicochemical properties of the spray-dried product were investigated. Five different responses were studied: (i) powder flow, (ii) yield, (iii) drug loading, (iv) particle size upon dilution (1:1000 *w*/*w*), and (v) the percentage of drug release at 60 min in simulated gastric fluid (500 mL buffer solution of pH 1.2).

### 2.6. Characterisation of the Solid B-SNEDDS

#### 2.6.1. Powder Flow: Angle of Repose (AoR) Measurements

The angle of repose (AoR) was determined according to the United State Pharmacopeia (<1174> Powder Flow) by using the fixed height funnel method (i.e., by measuring the cone height versus the base, formed by the powder falling through a plastic funnel placed 10 cm from the table surface until a stable cone was formed) [23]. AoR measurements were carried out by passing 500 mg of solid B-SNEDD in triplicate. After the powder was deposited on the surface, the height of the powder cone ($h$) and the radius of the base ($r$) were measured. The inclination of the cone created between the powder and the surface ($\alpha$, AoR) was calculated using the following equation:

$$\tan(\alpha) = \frac{h}{r} \quad (2)$$

#### 2.6.2. Particle Size Measurements

The particle size of both liquid B-SNEDDS and solid B-SNEDDS were determined at 25 °C after 1:1000 *w*/*w* dilution in deionised water. Solid B-SNEDDS were centrifuged (9000 rpm, 5 min) prior to measurements, in order to precipitate the insoluble carrier. The mean size ($n$ = 3) by volume (nm) was measured using a Microtrac Zetatrac (Microtac Inc., Montgomeryville, PA, USA), with an internal probe ranging from 0.0008 to 6.5 μm [24].

#### 2.6.3. Release Studies

Release studies of solid B-SNEDDS (500 mg) containing 10 mg of butenafine were performed at 37 °C for 1 h in simulated gastric fluid (SGF; 500 mL buffer solution of pH 1.2), followed by a second consecutive hour in simulated intestinal fluid (SIF; 400 mL buffer solution of pH 6.8) in a calibrated dissolution apparatus (Erweka type DT80, Erweka, Heusenstamm, Germany). The SGF and SIF were prepared as described in the USP [25]. Based on predicted water solubility values ($7.5 \times 10^{-5}$ mg/mL, according to the Drug Bank Database [26]), release studies were performed in non-sink conditions, with the aim of testing a relevant pharmacological drug dose (10 mg). At different time points (5, 10, 15, 30, 45, 60, 90, and 120 min), a 2 mL sample was withdrawn, filtered through a Millipore Millex PTFE membrane filter (0.45 μm), diluted 1:2 with acetonitrile, and injected in the HPLC. Butenafine concentration was quantified using an HPLC equipped with a Jasco PU-1580 pump, a Jasco AS-2050 Plus autosampler, and a Jasco UV-1575 UV-visible detector. Integration of the peaks was performed using the Borwin 1.5 software. Butenafine was separated on an Agilent Eclipse XDB-Phenyl reverse-phase column (250 mm × 4.6 mm, 5 μm). The mobile phase consisting of methanol/water (78:22 *v*/*v*) was pumped at 1.4 mL/min, and a sample injection volume of 20 μL was used. The column temperature was kept at 25 °C, and the detector was set at 220 nm. Butenafine concentrations were calculated from a linear regression calibration curve between 100.0–0.1 μg/mL.

#### 2.6.4. Drug Loading

Solid B-SNEDD formulations (10 mg) were dissolved in 0.5 mL of dimethyl sulfoxide (DMSO), vortexed for 5 min, and then diluted in acetonitrile (1:40 *v/v*) prior to drug quantification by the above-described HPLC method.

#### 2.6.5. Morphological Analysis

The morphology of solid B-SNEDDS was examined using a scanning electron microscope (JEOL JSM 6335F Ltd., Akishima, Japan) at 5 kV. The samples were fixed on a brass stub using double-sided adhesive tape, and vacuum-coated with gold (Au) for 180 s (coater: Q150R S, Quorum, Lewes, East Sussex, United Kingdom).

#### 2.6.6. Tabletting and Hardness

In order to investigate the compression of the solid B-SNEDDS without the addition of any other excipient, solid B-SNEDDS (500 mg) were compressed using a Perkin Elmer hydraulic press (Waltham, MA, USA) and a 13 mm punch, and die set under different pressures of 0.5, 1.0, 3.0 or 5.0 tonnes for 15 s. Hardness was undertaken according to the European Pharmacopeia [27], using a Pharma Test PTB 311 instrument (Pharma Test, Hainburg, Germany). Tablets ($n$ = 3) were individually evaluated, and the mean value of the force in Newtons (N) was reported.

### 2.7. *In Vitro Efficacy and Toxicity Studies*

#### 2.7.1. Parasites and Cell Lines

The parasites were kindly provided by Prof. Dr. Fernando Tobias Silveira from the cryobank of the Leishmaniasis Laboratory of Prof. Dr. Ralph Laison, Department of Parasitology, Ministry of Health, Evandro Chagas Institute (Belem, Para, Brazil). They were identified using monoclonal antibodies and isoenzyme electrophoretic profiles at the Leishmaniasis Laboratory of the Evandro Chagas Institute. The *Leishmania* species used was *L. (L.) infantum* (MHOM/BR/72/46). Parasites in a late log stage were used for all experiments. Parasites were maintained in Schneider's Medium (SigmaAldrich, Madrid, Spain), supplemented with 10% heat-inactivated fetal bovine serum, 50,000 IU/mL penicillin, and 50 μg/mL streptomycin.

BALB/c mice, 6 weeks old, were obtained from the Medical School of the University of São Paulo, Brazil, in order to collect peritoneal macrophages to perform the in vitro test against intracellular *Leishmania* amastigotes. To obtain the macrophages, the animals were anaesthetized with thiopental (1 mg/200 mL) and euthanized. This study was carried out in strict accordance with the recommendations detailed in the Guide for the Care and Use of Laboratory Animals of the Brazilian National Council of Animal Experimentation (http://www.cobea.org.br). The protocol was approved by the Ethics Committee of Animal Experiments of the Institutional Committee of Animal Care and Use at the Medical School of Sao Paulo University (CEUA-FMUSP number 098/17).

#### 2.7.2. In Vitro Promastigote Efficacy and Cytotoxicity

Promastigote forms of *L. (L.) infantum* ($2 \times 10^7$ promastigotes/well) were incubated in a 96-well culture plate in RPMI 1640 medium, with pH 4.2 and drugs in a range of 0.01 to 400 μM. The negative control group was cultivated in medium and vehicle solution (PBS plus 1% DMSO). The parasites were incubated for 48 h at 25 °C, and washed with 200 μL of sodium chloride 0.9% (*w/v*) three times with centrifugation at 3000 rpm for 10 min at 4 °C, followed by the addition of MTT (3-(4,5-dimethylthiazol-2-yl)-2,5-diphenyltetrazolium bromide) (9.6 μM). Four hours later, 50 μL of 10% sodium dodecyl sulphate (SDS) was added to each well. The plates were further incubated for 18 h and read in an ELISA reader (Labsystems Uniscience Multiskan EX, Miami, FL, USA) at 595 nm. Effective concentration 50% ($EC_{50}$) was estimated using Graph Pad Prism 5.0 software (GraphPad Software, San Diego, CA, USA).

Approximately $5 \times 10^5$ peritoneal macrophages from BALB/c mice were cultured in RPMI 1640 medium, with the drugs in a range of 0.01 to 400 μM, in 96-well plates. As a negative control, macrophages were cultivated with vehicle solution. After 48 h, cell viability was analysed by the MTT method. Cytotoxic concentration 50% ($CC_{50}$) was estimated with Graph Pad Prism 5.0 software.

The selectivity indexes (SI) were calculated using the ratio $CC_{50}/EC_{50}$ toward promastigote (SIp) or amastigote (SIa) forms. $EC_{50}$ represents the concentration of the formulation that produced a 50% reduction in parasites, while $CC_{50}$, represents the concentration of the formulation that produced a 50% reduction of cell viability in treated culture cells with respect to untreated ones.

#### 2.7.3. Macrophage Infection and Treatments

Peritoneal macrophages from BALB/c mice ($5 \times 10^5$ macrophages) were cultivated in round cover slips in a 24-well plate, followed by infection with *L. (L.) infantum* promastigotes at a ratio of 10 parasites per 1 peritoneal macrophage. Plates were incubated at 5% $CO_2$ at 37 °C. After 24 h of the initial infection, drugs were added at 25, 50, or 100 μM. Round cover slips from each experimental time point were dried at room temperature, fixed in methanol, and stained by Giemsa 5% (two drops). The infection index (II) was then estimated according to Passero et al. [28], using the following equation:

$$II = \%\ infected\ macrophages \times \frac{Internalized\ amastigotes}{Macrophages} \quad (3)$$

### *2.8. Statistical Analysis*

Statistical analyses were performed via a one-way ANOVA test using Minitab 15 (Minitab Ltd., Coventry, UK), followed by Tukey's test. Statistical significance was set at $p < 0.05$. Data was plotted using Origin X9 (Northampton, UK).

## 3. Results

### *3.1. Solubility Studies of Butenafine*

The solubility of butenafine was tested in four different excipients (Peceol, Capryol 90, Labrasol, and Labrafil M1944CS) that are commonly utilized in the development of lipid-based formulations and have been shown to possess efficacy against leishmaniasis [22] (Table 1). Butenafine was more soluble in excipients with oleic acid lipids or triglycerides, i.e., Peceol and Labrafil M 1944 CS, followed by lower HLB excipients such as Capryol 90. Due to the low miscibility between Peceol and Labrafil, and the higher butenafine solubility in Peceol, Peceol was chosen as the oil phase to be combined with Capryol and Labrasol for the pseudo-ternary diagram and identification of the optimal composition (Table 2) able to yield microemulsions (type II) upon aqueous dilutions.

**Table 1.** Solubility of butenafine in various vehicles at 25 °C ($n = 3$).

| Vehicle | Butenafine (mg/g) | HLB * |
|---|---|---|
| Peceol | 24.68 ± 0.21 | 1 |
| Capryol 90 | 15.79 ± 1.18 | 5 |
| Labrasol | 13.03 ± 3.94 | 12 |
| Labrafil M 1944 CS | 22.59 ± 1.66 | 9 |

* Hydrophilic-lipophilic balance of the vehicles; values obtained from Gatefosse website [29].

**Table 2.** Design of experiments (DoE) of liquid butenafine self-nanoemulsifying drug delivery systems (B-SNEDDS). Excipient quantities are expressed as a fraction, considering that the sum of all excipients was equal to 1 g. For each combination, the average particle size ($n = 3$) in numbers after dilution in de-ionised water (1:1000 *w/w*) was illustrated.

| Experiment | Labrasol | Capryol 90 | Peceol | Particle Size (nm) |
|---|---|---|---|---|
| 1 | 0.29 | 0.50 | 0.2 | 116 |
| 2 | 0 | 0.50 | 0.49 | 376 |
| 3 | 0 | 0.8 | 0.2 | 289 |
| 4 | 0.19 | 0.2 | 0.60 | 191 |
| 5 | 0.22 | 0.41 | 0.35 | 156 |
| 6 | 0.29 | 0.50 | 0.2 | 178 |
| 7 | 0.09 | 0.61 | 0.29 | 244 |
| 8 | 0 | 0.2 | 0.8 | 1735 |
| 9 | 0 | 0.50 | 0.49 | 410 |
| 10 | 0.39 | 0.34 | 0.26 | 107 |
| 11 | 0.4 | 0.2 | 0.4 | 101 |
| 12 | 0.12 | 0.35 | 0.52 | 167 |
| 13 | 0 | 0.2 | 0.8 | 1340 |
| 14 | 0 | 0.34 | 0.65 | 1180 |
| 15 | 0 | 0.8 | 0.2 | 352 |
| 16 | 0.19 | 0.2 | 0.60 | 141 |

### *3.2. Pseudo-Ternary Phase Diagrams and Preparation of Liquid B-SNEDDS Formulations*

A ternary phase diagram (Table 2 and Figure 1) was constructed to study the phase behaviour of oil/surfactants over the whole concentration range. Particle size expressed in numbers led to a better predictive model with a higher $R^2$ than the one using values expressed in volume. The blue region indicates the self-nanoemulsifying region, with a lower particle size. The particle size optimisation studies suggested two optimal excipient combinations with the following composition:

(1) Combination A: Peceol:Capryol:Labrasol mixed at 27:58:15 *w/w*
(2) Combination B: Peceol:Capryol:Labrasol mixed at 25:51:24 *w/w*

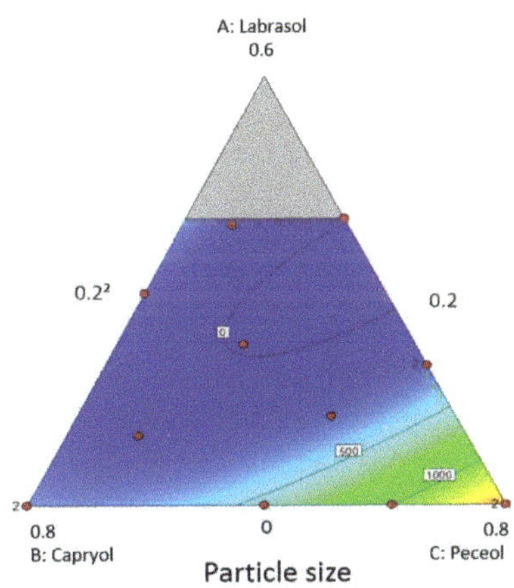


**Figure 1.** Pseudo-ternary phase diagram of SNEDDS. In the white rectangles is indicated the particle size of each corresponding composition. The blue areas indicate a lower particle size, while yellow and red areas indicate larger sizes.

Validation studies were performed by preparing the suggested optimal mixtures, followed by measuring the particle size. The resulted particle size was 159.7 nm for combination A, and 88.4 nm for combination B. Based on these results, combination B was selected for further development of solid-B-SNEDD formulations.

### 3.3. Yield of Solid B-SNEDDS

Four different solid B-SNEDDS were manufactured utilising the vehicle mixture (combination B) above described. Two different hydrophobic, fumed silica carriers (Aerosil 200 or R972) were combined in two ratios (1:2 and 1:3 *w/w*), with the optimised liquid B-SNEDDS and the resulting four formulations coded as F1 to F4, respectively (Table 3). When SNEDDS were adsorbed on Aerosil 200 at a 1:2 ratio (F1), a higher yield was obtained, which was able to carry higher amounts of liquid B-SNEDDS (1:3 ratio). Hydrophobic fumed silica resulted in a poorer yield, indicating that SNEDDS adsorbed easier towards a hydrophilic surface, suggesting a core shell particle structure for liquid B-SNEDDS, with Labrasol being orientated towards the surface, and likely to interact more strongly with hydrophilic silica surfaces. However, this stronger interaction of labrasol with hydrophilic silicas has been linked previously to poorer disintegration times, which should be taken into account later when manufacturing solid dosage forms, such as tablets [30].

**Table 3.** Characterisation of the solid B-SNEDDS. Results are reported as mean ± SD ($n$ = 3).

| Code | Aerosil Type | Carrier: B-SNEDDS Ratio (*w/w*) | Yield (%) | Angle of Repose (°) | Drug Loading (%) | Particle Size (nm) | Butenafine Released at 60 min in SGF (%) |
|---|---|---|---|---|---|---|---|
| F1 | 200 | 1:2 | 71.1 ± 1.6 | 19.1 ± 5.4 | 94.1 ± 0.1 | 119.8 ± 2.3 | 45 ± 2.0 |
| F2 | 200 | 1:3 | 66.5 ±10.2 | 15.9 ± 3.0 | 96.5 ± 2.4 | 25.9 ± 1.2 | 28 ± 5.0 |
| F3 | R972 | 1:2 | 32.9 ± 1.3 | 20.6 ± 0.2 | 85.5 ± 4.6 | 279 ± 8.7 | 18 ± 4.0 |
| F4 | R972 | 1:3 | 47.4 ± 22.6 | 8.7 ± 1.5 | 97.0 ± 2.1 | 1 * | 20 ± 0.2 |

* Particle size measurements of F4 were registered as 1 nm, indicating that particles were not stable upon dilution.

All solid B-SNEDDS had an AoR below 20°, indicative of excellent flow properties [31,32] with near maximal drug loading (Table 3). Particle size measurements confirmed microemulsion regions with a size well below 300 nm. F4 yielded particles that either had not been able to re-form after adsorption on silica, or were unstable after the 1:1000 dilution, and thus were considered not appropriate for further development.

### 3.4. Release Studies

Hydrophilic fumed silica particles indicated a greater and faster release of butenafine nanoemulsions type II (F1 > F2 > F3~F4) (Figure 2). Increasing the pH from 1.2 to 6.8 to mimic the intestinal pH resulted in a sharp precipitation, which can be linked with a spring–parachute effect. This can explain the low absorption of butenafine orally [21], as that only allows for absorption in the upper part of the gastrointestinal tract [33,34]. Butenafine is more soluble in acidic pH as a hydrochloride salt, while its base (pKa: 9.23) has limited water solubility (<100 ng/mL) [35]. Formulation F1 allows for higher solubilised levels up to 90 min ($p < 0.05$; one-way ANOVA), which is critical for enhancing oral absorption.

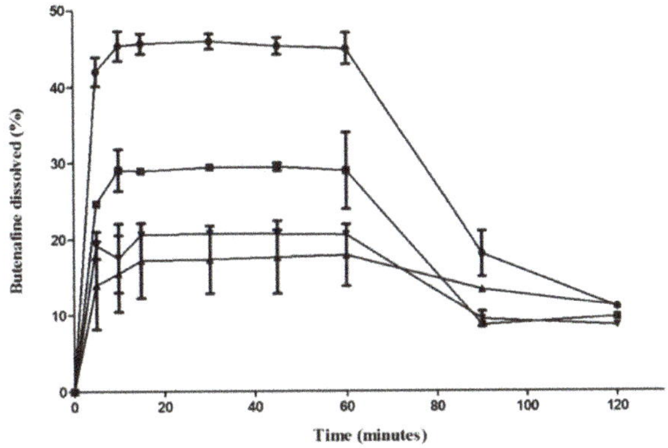


**Figure 2.** Release of solid B-SNEDDS formulations in simulated gastric fluid (SGF) (pH 1.2, first 60 min) and simulated intestinal fluid (SIF) (pH 6.8) thereafter (mean % ± SD). -●- F1, -■-F2, -▲- F3, -▼- F4.

## 3.5. Morphological Analysis

The morphology of the four solid B-SNEDDS formulations was observed by scanning electron microscopy (Figure 3). At higher magnification, the large surface porosity of both silicon dioxide carriers can be observed. This explains the high amount of liquid SNEDDS that is able to be adsorbed on to the solid carriers, up to three times their own weight. However, agglomeration between carrier particles was more noticeable in those formulations containing a higher ratio of liquid SNEDDS (F2 and F4), which can explain the hindered release observed. In the F3 and F4, both containing the most hydrophobic carrier (Aerosil R972), it was observed that even at the lower ratio of liquid-SNEDDS (1:2 *w:w*), the aggregation between particles is still evident to some extent.

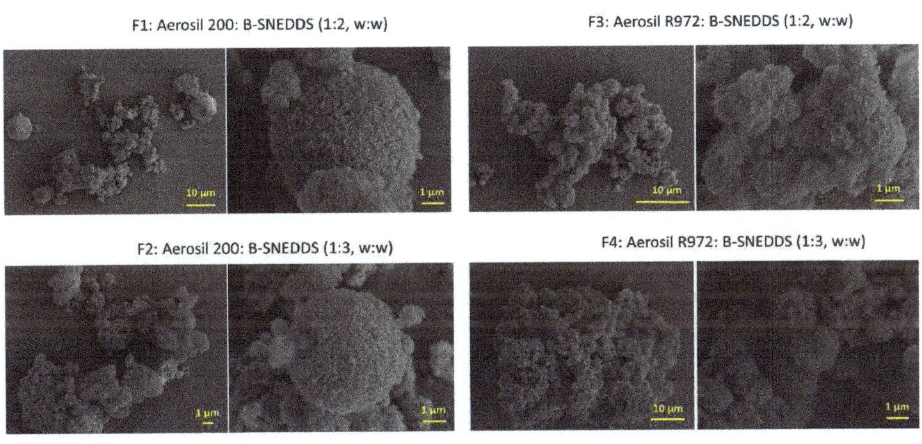


**Figure 3.** Scanning electron microscope (SEM) micrographs of solid B-SNEDDS F1, F2, F3, and F4 formulations at two different magnifications.

## 3.6. Tableting and Hardness

Amongst all the solid B-SNEDDS, only the F1 formulation was able to be compacted with adequate hardness without the incorporation of other excipients. We should bear in mind that the addition of other excipients will further dilute the butenafine dose per tablet size, which currently is 10 mg/500 mg tablet in powder form (Table 4). The addition of microcrystalline cellulose is likely to allow other solid SNEDDS to be formulated with appropriate hardness.

**Table 4.** Hardness of the F1 solid B-SNEDDS formulation. Hardness expressed as minimum and maximum values.

| Compaction Pressure (kN) | Hardness (N) | Dimension (mm) |
|---|---|---|
| 9806.65 | 15.6–18.4 | 12.91 × 3.17 |
| 19,613.3 | 7.7–20.1 | 12.91 × 3.17 |

### 3.7. In Vitro Efficacy and Cytotoxicity

Miltefosine was active against both promastigote and amastigote forms of *L. (L.) infantum*, exhibiting an $EC_{50}$ of 17.9 ± 0.9 and 13.7 ± 0.7μM, respectively, as well as a mild cytotoxicity ($CC_{50}$ of 126.3 ± 3.5 μM) leading to a selectivity index towards promastigotes forms (SIp) of 7.0 and a selectivity index towards amastigotes forms (SIa) of 9.2 (Table 5). These values are similar to previous documented studies [36]. Free butenafine showed lower efficacy than miltefosine (Table 5), with an SI close to the unit indicating poor selectivity for the parasites over the macrophages. Amongst the four solid B-SNEDDS, F1 and F4 showed better efficacy against promastigotes ($p < 0.05$), while Aerosil R972 formulations (F3 and F4) demonstrated lower cytotoxicity and higher efficacy against amastigotes ($p < 0.05$), but whether this is driven by the excipient is not clear, and further studies are needed. In any case, except for F2, formulating butenafine as SNEDDS has improved its selectivity index, making it less toxic and more active than the free drug.

**Table 5.** Antileishmanial activity of solid B-SNEDDS formulations and butenafine were assayed against promastigote and amastigote forms of *L. (L.) infantum*. Cytotoxicity was analysed using peritoneal macrophages from BALB/c mice. $EC_{50}$ represents the concentration of the formulation that produced a 50% reduction in parasites, while $CC_{50}$ represents the concentration of the formulation that produced a 50% reduction of cell viability in treated culture cells with respect to untreated ones. SIp: selectivity index towards promastigotes forms; SIa: selectivity index towards amastigotes forms.

| Formulation | $EC_{50}$ (μM) Promastigote | $CC_{50}$ (μM) Macrophage | $EC_{50}$ (μM) Amastigote | SIp | SIa |
|---|---|---|---|---|---|
| F1 | 76.5 ± 3.0 * | 225.8 ± 3.2 | 164.4 ± 3.4 | 2.9 | 1.3 |
| F2 | 93.1 ± 3.0 | 127.2 ± 3.0 | 702.5 ± 3.2 | 1.3 | 0.1 |
| F3 | 101.7 ± 3.2 | ≥300.0 | 86.4 ± 2.9 * | ≥3.0 | ≥3.6 |
| F4 | 73.9 ± 3.2 * | 233.1 ± 3.0 | 27.0 ± 3.4 * | 3.1 | 8.6 |
| Butenafine | 99.8 ± 3.1 | 109.3 ± 3.5 | 118.4 ± 3.6 | 1.0 | 0.9 |
| Miltefosine | 17.9 ± 0.9 | 126.3 ± 3.5 | 13.7 ± 0.7 | 7.0 | 9.2 |

* $p \leq 0.05$ compared to Butenafine.

## 4. Discussion

Butenafine, a benzylamine derivative structurally similar to terbinafine, possesses antifungal activity, attributed to its ability to directly cause damage on fungal cell membranes by disrupting the early stages of ergosterol biosynthesis via inhibition of the enzyme squalene epoxidase [37]. This enzyme converts squalene to lanosterol, and leads to the accumulation of squalene [38]. Inhibition of squalene epoxidase suppresses the biosynthesis of ergosterol, an essential lipid of fungal and *Leishmania* cell membranes [39].

Therefore, butenafine is known for its antifungal effects in infections caused by *Tinea pedis*, *Tinea corporis*, and *Tinea cruris* [40]. We have recently demonstrated the leishmanicidal effect of butenafine against promastigote and amastigote forms of *L. (L.) amazonensis* and *L. (V.) braziliensis* [20]. Considering the difficulties related to the treatment of leishmaniasis, such as drug relapses, toxicity, hospitalization, and parenteral administration, the development of an oral, safe, multispecies, and effective medicament could revolutionise VL treatment. Butenafine is a multispecies drug, able to be effective against a range of parasite strains, such as *L. (L.) amazonensis*, *L. (V.) braziliensis*, and now in this work, it has been demonstrated that is also active against *L infantum*. However, due to its poor water solubility and precipitation in the gastrointestinal tract, no oral formulations of butenafine have been licensed.

In order to pave the way for the market of oral medicines against VL, the selection of excipients that can promote oral bioavailability and safety, as well as elicit a synergistic effect against Leishmania parasites without compromising safety is critical. Peceol, Labrasol, and Capryol 90 have been selected as GRAS excipients to formulate butenafine SNEDDS, able to enhance the drug solubilisation capacity in the gastrointestinal tract and oral bioavailability [41]. Also, due to the proven efficacy of those selected excipients against several strains of Leishmania parasites, a synergistic effect with butenafine can be expected [22].

Pre-formulation studies have indicated that butenafine is more soluble in C18 lipids and amphiphiles, and these were thus selected for forming the oil component of SNEDDS (Peceol). We decided to combine these with the medium, as well as short chain fatty acids esters (Capryol 90) and triglycerides (Labrasol) that we have recently shown to possess antileishmanial activity. The high reported activity of lauric acid and labrasol against *Leishmania* parasites can be attributed to their ability to selectively permeate the cell membrane of parasites, resulting in rapid and considerable membrane damage and the loss of cellular potassium and magnesium [22]. SNEDDS were optimised towards a minimised droplet particle size (~100 nm) for several reasons. Nanoemulsions (type II microemulsions) are more stable than microemulsions, in terms of droplet flocculation and coalescence (Ostwald ripening) [42]. The smaller the particle size, the higher the permeability across the intestinal mucus brush border layer and cell membranes [43]. Additionally, small droplet sizes (~90 nm) provide a large interfacial surface area for drug release and absorption across the intestinal cells [44]. Recent reports suggest that a particle size between 100–500 nm is optimal for lymphatic uptake via the gastrointestinal lymphatic system, but at a slower rate than particles sized 50–100 nm [45,46].

In order to reduce costs, increase the chemical stability, and improve patient compliance, B-SNEDDS was transformed into a solid free-flowing powder by spray-drying, which can be directly compressed. Fumed amorphous silica (Aerosil) are known to possess enhanced surface area for adsorption (Aerosil 200: 200 ± 25 $m^2/g$ and Aerosil R972: 110 ± 20 $m^2/g$). Both silicas were able to adsorb three times their own weight in SNEDDS, resulting in powders with excellent flow properties, as indicated by AoR studies (<20°) [30]. The higher porosity and hydrophilic surface of Aerosil 200 carriers explain the higher drug release and yield, respectively. Loading was higher with Aerosil 200, which indicates a core-shell morphology of the SNEDDS droplet, with polyethylene glycol chains stabilising the droplet surface and enabling higher interactions with the hydrophilic surface of the silica [47,48]. Aerosil R972 carriers hinder the release of butenafine, possibly due to stronger hydrophobic interactions; however, this is likely to limit oral bioavailability. Additionally, particles prepared with Aerosil R972, with high SNEDDS loading when dispersed in water, are unable to yield an emulsion, indicating that droplets are not able to re-form or are unable to maintain their morphology, and become unstable after a 1:1000 dilution. The latter, combined with limited release from these solid B-SNEDDS, makes them less likely to be able to enhance oral bioavailability. Interestingly, though, these particles have a significantly impact on the in vitro efficacy against *Leishmania* promastigotes and amastigotes. Even though butenafine release from F3 and F4 formulations was low at physiological pH, these formulations possessed enhanced efficacy and selectivity, with no higher cytotoxicity on macrophages. Enhanced surface hydrophobicity can trigger macrophage uptake compared to particles with a hydrophilic surface [49]. However, when the results are taken together, formulation F1 is more likely to be a promising solid nanomedicine for VL, as it maintains characteristics for enhanced solubilisation and oral uptake, as well as presented antileishmanial activity in *L. (L.) infantum* promastigotes and amastigotes.

Limited studies are available in the literature for understanding the ability of solid carriers to load SNEDDS. Preparing a conventional solid dosage form of SNEDDS remains challenging, as tablets can be friable if a low compression force is used, or allow for SNEDDS leakage if a high compression force is used. Tablets prepared with Labrasol and silica (Neusilin US2, a synthetic, amorphous form of magnesium aluminometasilicate) have demonstrated an enhanced disintegration time and the need for a disintegrant [32]. Our solid B-SNEDDS were able to load high SNEDDS quantities, while being free-flowing and easily compacted into the appropriate hardness tablets. Smith et al. [12] have demonstrated that liquid SNEDDS can be transformed into solid tablets with good release properties when the ratio between liquid to solid was 1:3. In our work, we have shown that even a much greater ratio of liquid to solid (2:1) can still be compacted without SNEDDS leakage and maintain a homogenous matrix for compaction. However, further studies are required to investigate the manufacturing of these free-flowing powders into tablets that meet the USP Pharmacopeia requirements.

## 5. Conclusions

Manufacturing of cost-effective solid dosage forms of reformulated drugs is an ideal strategy to speed up the development of novel medicines for VL and other neglected diseases. We have demonstrated for the first time the efficacy of butenafine for VL, while developing an easily scalable preparation from GRAS excipients of solid dosage forms of butenafine, able to maintain its solubilisation capacity in the gastrointestinal tract. SNEDDS were optimised for drug loading (30 mg/g) and particle size, and we have demonstrated the ability of a microemulsion to be formed after release from fumed silica particles. Solid SNEDDS demonstrated excellent flow properties, and were able to be compressed in adequate hardness tablets. They also demonstrated antileishmanial activity in *L. (L.) infantum* promastigotes and amastigotes, which indicated their potential as solid nanomedicines for the treatment of VL.

**Author Contributions:** Conceptualization, A.L. and D.R.S.; methodology, A.B.-S., G.F.R., R.F.-G., L.F.P., M.D.L., and D.R.S.; software, A.B.-S.; validation, A.B.-S.; formal analysis, A.B.S, G.F.R., R.F.-G., and D.R.S.; resources, M.D.L., F.B.-F., L.F.P., and D.R.S.; data curation, A.B.-S.; writing—original draft preparation, A.B.-S.; writing—review and editing, A.B.-S., L.F.P., A.L., and D.R.S.; supervision, M.D.L, L.F.P., and D.R.S.; project administration, M.D.L. and F.B.-F.; funding acquisition, M.D.L. and F.B.-F.

**Funding:** This research was funded by the Unión Iberoamericana de Universidades (ENF03-2017) allowing student mobility exchange between UCM and USP and by Fundação de Amparo à Pesquisa do Estado de São Paulo (FAPESP processes: 2016/00468-0 and 2017/09405-4).

**Conflicts of Interest:** The authors declare no conflict of interest.

## References

1. Araújo-Santos, T.; Prates, D.B.; Andrade, B.B.; Nascimento, D.O.; Clarêncio, J.; Entringer, P.F.; Carneiro, A.B.; Silva-Neto, M.A.; Miranda, J.C.; Brodskyn, C.I.; et al. Lutzomyia longipalpis saliva triggers lipid body formation and prostaglandin $E_2$ production in murine macrophages. *PLoS Negl. Trop. Dis.* **2010**, *4*, e873. [CrossRef] [PubMed]
2. World Health Organization (WHO). Leishmaniasis Disease and Epidemiology. Available online: http://www.who.int/leishmaniasis/epidemic/response_more/en/index.html (accessed on 18 February 2019).
3. Rittig, M.G.; Bogdan, C. Leishmania-Host-cell Interaction: Complexities and Alternative Views. *Parasitol. Today* **2000**, *16*, 292–297. [CrossRef]
4. Laurenti, M.D.; Da Matta, V.L.R.; Pernichelli, T.; Secundino, N.F.C.; Pinto, L.C.; Corbett, C.E.P.; Pimenta, P.P.F. Effects of Salivary Gland Homogenate from Wild-Caught and Laboratory-Reared Lutzomyia longipalpison the Evolution and Immunomodulation of Leishmania (Leishmania) amazonensis Infection. *Scand. J. Immunol.* **2009**, *70*, 389–395. [CrossRef] [PubMed]
5. Laison, R.; Shaw, J.J. New world Leishmaniasis—The Neotropical Leishmania species. In *Topley & Wilson. Microbiology and Microbial Infections*, 9th ed.; Feg Cox: London, UK, 1988.
6. Silveira, F.T.; Lainson, R.; Crescente, J.Â.; De Souza, A.A.; Campos, M.B.; Gomes, C.M.; Laurenti, M.D.; Corbett, C.E. A prospective study on the dynamics of the clinical and immunological evolution of human Leishmania (L.) infantum chagasi infection in the Brazilian Amazon region. *Trans. R. Soc. Trop. Med. Hyg.* **2010**, *104*, 529–535. [CrossRef] [PubMed]
7. Dantas-Torres, F. Leishmania infantum versus Leishmania chagasi: Do not forget the law of priority. *Memórias do Instituto Oswaldo Cruz* **2006**, *101*, 117–118. [CrossRef]

8. Barak, E.; Amin-Spector, S.; Gerliak, E.; Goyard, S.; Holland, N.; Zilberstein, D. Differentiation of Leishmania donovani in host-free system: Analysis of signal perception and response. *Mol. Biochem. Parasitol.* **2005**, *141*, 99–108. [CrossRef]
9. Santos, D.O.; Coutinho, C.E.R.; Madeira, M.F.; Bottino, C.G.; Vieira, R.T.; Nascimento, S.B.; Bernardino, A.M.R.; Bourguignon, S.C.; Côrte-Real, S.; Pinho, R.T.; et al. Leishmaniasis treatment—A challenge that remains: A review. *Parasitol. Res.* **2008**, *103*, 1–10. [CrossRef]
10. Kaur, G.; Rajput, B. Comparative Analysis of the Omics Technologies Used to Study Antimonial, Amphotericin B, and Pentamidine Resistance in Leishmania. *J. Parasitol. Res.* **2014**, *2014*, 1–11. [CrossRef]
11. Serrano, D.R.; Lalatsa, A. Oral amphotericin B: The journey from bench to market. *J. Drug Deliv. Sci. Technol.* **2017**, *42*, 75–83. [CrossRef]
12. Smith, L.; Serrano, D.R.; Mauger, M.; Bolás-Fernández, F.; Dea-Ayuela, M.A.; Lalatsa, A. Orally Bioavailable and Effective Buparvaquone Lipid-Based Nanomedicines for Visceral Leishmaniasis. *Mol. Pharm.* **2018**, *15*, 2570–2583. [CrossRef]
13. Fernández, O.L.; Díaz-Toro, Y.; Ovalle, C.; Valderrama, L.; Muvdi, S.; Rodriguez, I.; Gomez, M.A.; Saravia, N.G. Miltefosine and Antimonial Drug Susceptibility of Leishmania Viannia Species and Populations in Regions of High Transmission in Colombia. *PLoS Negl. Trop. Dis.* **2014**, *8*, e2871. [CrossRef]
14. Dorlo, T.P.C.; Balasegaram, M.; Beijnen, J.H.; De Vries, P.J. Miltefosine: A review of its pharmacology and therapeutic efficacy in the treatment of leishmaniasis. *J. Antimicrob. Chemother.* **2012**, *67*, 2576–2597. [CrossRef]
15. Pérez-Victoria, F.J.; Sánchez-Cañete, M.P.; Seifert, K.; Croft, S.L.; Sundar, S.; Castanys, S.; Gamarro, F. Mechanisms of experimental resistance of Leishmania to miltefosine: Implications for clinical use. *Drug Resist. Updates* **2006**, *9*, 26–39. [CrossRef]
16. Drugs for Neglected Diseases initiative: DNDi. Available online: https://www.dndi.org/diseases-projects/leishmaniasis/tpp-vl (accessed on 19 January 2018).
17. Graul, A.; Sorbera, L.; Pina, P.; Tell, M.; Cruces, E.; Rosa, E.; Stringer, M.; Castaner, R.; Revel, L. The year's new drugs & biologics–2009. *Drug News Perspect.* **2010**, *23*, 7. [CrossRef]
18. Tobinick, E.L. The value of drug repositioning in the current pharmaceutical market. *Drug News Perspect.* **2009**, *22*, 119. [CrossRef]
19. Voss, A.; Soto, J.; Toledo, J.; Nicholls, R.S.; Padilla, J.; Engel, J.; Fischer, C.; Gutierrez, P.; Berman, J. Treatment of American Cutaneous Leishmaniasis with Miltefosine, an Oral Agent. *Clin. Infect. Dis.* **2001**, *33*, e57–e61.
20. Bezerra-Souza, A.; Yamamoto, E.S.; Laurenti, M.D.; Ribeiro, S.P.; Passero, L.F.D. The antifungal compound 25butenafine eliminates promastigote and amastigote forms of Leishmania (Leishmania) amazonensis and Leishmania (Viannia) braziliensis. *Parasitol. Int.* **2016**, *65*, 702–707. [CrossRef]
21. Food and Drug Administration (FDA). Review and Evaluation of Pharmacology/Toxicology Data O Fbutenafine, NDA 21-307 (000)/09-29-2000. Available online: https://www.accessdata.fda.gov/drugsatfda_docs/nda/2001/21-307_Lotrimin_pharmr.pdf (accessed on 24 June 2019).
22. Serrano, D.R.; Lalatsa, A.; Dea-Ayuela, M.A. Engineering Synergistically Active and Bioavailable Cost-effective Medicines for Neglected Tropical Diseases; The Role of Excipients. *Curr. Top. Med. Chem.* **2017**, *17*. [CrossRef]
23. Lavoie, F.; Cartilier, L.; Thibert, R. New Methods Characterizing Avalanche Behavior to Determine Powder Flow. *Pharm. Res.* **2002**, *19*, 887–893. [CrossRef]
24. Betatek Inc. *Instrumentation Superbly Supported: Microtrac Zetatrac Nanotechnology, Particle Size and Charge Measurement Analyzer*; Betatek Inc.: Toronto, ON, Canada, 2012.
25. United States Pharmacopeia (USP). *USP 41-NF36*; <711> Dissolution Studies; United States Pharmacopeial Convention: Rockville, MD, USA, 2008.
26. Drug Bank Database. Available online: https://www.drugbank.ca/drugs/DB01091 (accessed on 15 July 2019).
27. European Pharmacopeia, 9th ed. Available online: https://www.edqm.eu/en/european-pharmacopoeia-ph-eur-9th-edition. (accessed on 19 July 2018).
28. Passero, L.F.D.; Sacomori, J.V.; Tomokane, T.Y.; Corbett, C.E.P.; Da Silveira, F.T.; Laurenti, M.D. Ex vivo and in vivo biological behavior of Leishmania (Viannia) shawi. *Parasitol. Res.* **2009**, *105*, 1741–1747. [CrossRef]
29. Gattefossé: Excipients for Solubility and Bioavailability Enhancement. Available online: https://www.gattefosse.com/excipients-for-solubility-and-bioavailability-enhancement (accessed on 18 February 2019).

30. Maher, S.; Brayden, D.J.; Casettari, L.; Illum, L. Application of Permeation Enhancers in Oral Delivery of Macromolecules: An Update. *Pharmaceutics* **2019**, *11*, 41. [CrossRef]
31. United States Pharmacopeia (USP). *USP 41-NF36*; <1174> Powder Flow; United States Pharmacopeial Convention: Rockville, MD, USA, 2006.
32. Goh, H.P.; Heng, P.W.S.; Liew, C.V. Comparative evaluation of powder flow parameters with reference to particle size and shape. *Int. J. Pharm.* **2018**, *547*, 133–141. [CrossRef]
33. Porter, C.J.H.; Pouton, C.W.; Cuine, J.F.; Charman, W.N. Enhancing intestinal drug solubilization using lipid-based delivery systems. *Adv. Drug Deliv. Rev.* **2008**, *60*, 673–691. [CrossRef]
34. Humberstone, A.J.; Charman, W.N. Lipid-based vehicles for the oral delivery of poorly water soluble drugs. *Adv. Drug Deliv. Rev.* **1997**, *25*, 103–128. [CrossRef]
35. Drugbank. Butenafine. Available online: https://www.drugbank.ca/drugs/DB01091 (accessed on 29 April 2019).
36. Seifert, K.; Escobar, P.; Croft, S.L. In vitro activity of anti-leishmanial drugs against Leishmania donovani is host cell dependent. *J. Antimicrob. Chemother.* **2010**, *65*, 508–511. [CrossRef]
37. Stütz, A. Synthesis and Structure—Activity Correlations within Allylamine Antimycotics. *Ann. N. Y. Acad. Sci.* **1988**, *544*, 46–62. [CrossRef]
38. Ryder, N.S. Terbinafine: Mode of action and properties of the squalene epoxidase inhibition. *Br. J. Dermatol.* **1992**, *126*, 2–7. [CrossRef]
39. Roberts, C.W.; McLeod, R.; Rice, D.W.; Ginger, M.; Chance, M.L.; Goad, L.J. Fatty acid and sterol metabolism: Potential antimicrobial targets in apicomplexan and trypanosomatid parasitic protozoa. *Mol. Biochem. Parasitol.* **2003**, *126*, 129–142. [CrossRef]
40. Singal, A. Butenafine and superficial mycoses: Current status. *Expert Opin. Drug Metab. Toxicol.* **2008**, *4*, 999–1005. [CrossRef]
41. Tatham, L.M.; Rannard, S.P.; Owen, A. Nanoformulation strategies for the enhanced oral bioavailability of antiretroviral therapeutics. *Ther. Deliv.* **2015**, *6*, 469–490. [CrossRef]
42. Anton, N.; Vandamme, T.F. Nano-emulsions and Micro-emulsions: Clarifications of the Critical Differences. *Pharm. Res.* **2011**, *28*, 978–985. [CrossRef]
43. Leonaviciute, G.; Adamovic, N.T.; Lam, H.T.; Rohrer, J.; Partenhauser, A.; Bernkop-Schnürch, A. Self-emulsifying drug delivery systems (SEDDS): Proof-of-concept how to make them mucoadhesive. *Eur. J. Pharm. Biopharm.* **2017**, *112*, 51–57. [CrossRef]
44. Imada, C.; Takahashi, T.; Kuramoto, M.; Masuda, K.; Ogawara, K.; Sato, A.; Wataya, Y.; Kim, H.S.; Higaki, K. Improvement of Oral Bioavailability of N-251, a Novel Antimalarial Drug, by Increasing Lymphatic Transport with Long-Chain Fatty Acid-Based Self-Nanoemulsifying Drug Delivery System. *Pharm. Res.* **2015**, *32*, 2595–2608. [CrossRef]
45. Khan, A.A.; Mudassir, J.; Mohtar, N.; Darwis, Y. Advanced drug delivery to the lymphatic system: Lipid-based nanoformulations. *Int. J. Nanomed.* **2013**, *8*, 2733–2744. [CrossRef]
46. Gosh, S.; Roy, T. Nanoparticulate drug-delivery systems: Lymphatic uptake and its gastrointestinal application. *J. Appl. Pharm. Sci.* **2014**, *4*, 123–130.
47. Elnaggar, Y.S.; El-Massik, M.A.; Abdallah, O.Y. Self-nanoemulsifying drug delivery systems of tamoxifen citrate: Design and optimization. *Int. J. Pharm.* **2009**, *380*, 133–141. [CrossRef]
48. Shen, S.; Wu, Y.; Liu, Y.; Wu, D. High drug-loading nanomedicines: Progress, current status, and prospects. *Int. J. Nanomed.* **2017**, *12*, 4085–4109. [CrossRef]
49. Serrano, D.R.; Hernández, L.; Fleire, L.; González-Alvarez, I.; Montoya, A.; Ballesteros, M.P.; Dea-Ayuela, M.A.; Miró, G.; Bolás-Fernández, F.; Torrado, J.J. Hemolytic and pharmacokinetic studies of liposomal and particulate amphotericin B formulations. *Int. J. Pharm.* **2013**, *447*, 38–46. [CrossRef]

*pharmaceutics*

Article

# Designing Fast-Dissolving Orodispersible Films of Amphotericin B for Oropharyngeal Candidiasis

**Dolores R. Serrano [1,2], Raquel Fernandez-Garcia [1], Marta Mele [3], Anne Marie Healy [4] and Aikaterini Lalatsa [3,*]**

1 Department of Pharmaceutics and Food Technology, School of Pharmacy, Universidad Complutense de Madrid, Plaza Ramon y Cajal s/n, 28040 Madrid, Spain
2 Instituto Universitario de Farmacia Industrial (IUFI), School of Pharmacy, Universidad Complutense de Madrid, Avenida Complutense, 28040 Madrid, Spain
3 School of Pharmacy and Biomedical Sciences, University of Portsmouth, St. Michael's Building, White Swan Road, Portsmouth PO1 2DT, UK
4 Synthesis and Solid State Pharmaceutical Centre, School of Pharmacy and Pharmaceutical Sciences, Trinity College Dublin, Dublin 2, Ireland
* Correspondence: katerina.lalatsa@port.ac.uk; Tel.: +44-023-9284-3929

Received: 7 June 2019; Accepted: 22 July 2019; Published: 1 August 2019

**Abstract:** Amphotericin B possesses high activity against *Candida* spp. with low risk of resistance. However, Amphotericin B's high molecular weight compared to other antifungal drugs, such as miconazole and clotrimazole, and poor water solubility hampers its efficacy at the physiological conditions of the oropharyngeal cavity (saliva pH, limited volume for dissolution) and thereby limits its clinical use in oropharyngeal candidiasis. We have prepared fast-dissolving orodispersible films with high loading (1% *w/w*) using solvent casting that enables amphotericin B to remain solubilised in saliva in equilibrium between the monomeric and dimeric states, and able to produce a local antifungal effect. Optimisation of the amphotericin B-loaded orodispersible films was achieved by quality by design studies combining dextran and/or maltodextrin as dextrose-derived-polymer film formers with cellulose-derived film formers (hydroxypropylmethyl/hydroxypropyl cellulose in a 1:4 weight ratio), sorbitol for taste masking, microcrystalline cellulose (Avicel 200) or microcrystalline cellulose-carboxymethylcellulose sodium (Avicel CL-611) for enhancing the mechanical strength of the film, and polyethylene glycol 400 and glycerol (1:1 *w/w*) as plasticizers. The optimised amphotericin B orodispersible films (containing 1% AmB, 25% dextran, 25% maltodextrin, 5% sorbitol, 10% Avicel 200, 10% polyethylene glycol 400, 10% glycerol, 3% hydroxypropylmethyl cellulose acetate succinate, 12% hydroxypropyl cellulose) possessed a fast disintegration time (60 ± 3 s), quick release in artificial saliva (>80% in 10 min), high burst strength (2190 mN mm) and high efficacy against several *Candida* spp. (*C. albicans*, *C. parapsilosis* and *C. krusei*) (>15 mm inhibition halo). Amphotericin B orodispersible films are stable for two weeks at room temperature (25 °C) and up to 1 year in the fridge. Although further toxicological and in vivo efficacy studies are required, this novel Amphotericin B orodispersible films is a promising, physicochemically stable formulation with potential wide application in clinical practice, especially for immunocompromised patients suffering from oropharyngeal candidiasis.

**Keywords:** orodispersible films; fast-dissolving films; micelles; amphotericin B; fungal infections

## 1. Introduction

Fungal infections of the oral cavity are opportunistic, usually caused by *Candida albicans*, and occur more frequently in patients that are immunocompromised (e.g., HIV, cancer patients), diabetics or having predisposing factors such as antibiotic and corticosteroid therapy, poor buccal hygiene and ill-fitted dentures. Local therapy of oral and pharyngeal candidiasis is desirable as it avoids adverse effects linked to systemic antifungal use [1]. A major challenge to effective local treatments remains the low volume for dissolution and need for rapid permeability of the oropharyngeal cavity in order to deliver adequate drug concentrations for local action. The majority of formulations rely on oral suspensions due to the poor aqueous solubility of most antifungals that are swished around in the mouth for a few seconds, gargled, and swallowed or spat out. This short contact time with the oral mucosa requires a readily available and solubilised drug to exert an antifungal effect.

Amphotericin B (AmB) is a broad spectrum antifungal effective in the nanomolar range ($IC_{50}$ of 0.25–1 μg $mL^{-1}$). Compared to azoles, there is a low frequency of *C. albicans* strains that are resistant to AmB [2]. The poor aqueous solubility of AmB (BCS Class IV) makes it difficult to solubilise in an adequate concentration in the small volume of saliva that is available in the oral cavity (1 mL) [3]. AmB is only commercialised as lyophilised formulations (micellar or lipidic nanoparticulate formulations) that are stable after reconstitution in aqueous media for a duration shorter than 24 h even when refrigerated [4,5] resulting in wastage, while poor patient compliance is an issue as formulations are not palatable. Taste masked orodispersable films (ODFs) can be potential solid dosage forms to deliver AmB in safe and efficient systems for the treatment of oropharyngeal fungal infections. Compared to other solid formulations, such as tablets and capsules, ODFs benefit from better patient compliance due to the ease of administration to dysphagic, paediatric and geriatric patients, without the need for water, and a rapid onset of action [6,7]. ODFs have shown better performance than semisolid formulations such as gels, because they can be easily transported, allow for accurate dosing and possess superior chemical and physical stability when packed appropriately, which can be important for unstable drugs in aqueous media such as AmB [8,9]. From an industrial and clinical point of view, ODFs would be more cost-effective formulations compared to parenteral AmB formulations, as they do not require sterilisation and lyophilisation, avoiding wastage.

The hypothesis underpinning this work is that AmB-loaded ODFs prepared using GRAS (Generally Regarded as Safe) excipients would enhance the current therapies available for the treatment of fungal infections in the oral and pharyngeal cavities. However, to the best of our knowledge, this has not been achieved to date due to the physicochemical challenges when formulating this drug. Here, we present an optimised fast disintegrating ODF of AmB with improved stability, loaded with high amounts of drug and designed to ensure AmB solubility in small volumes of saliva, while being taste masked and locally effective. To ensure AmB solubility at the physiological pH of the oropharyngeal cavity, we entrapped AmB in sodium deoxycholate micelles that were then embedded within the ODF. Sodium deoxycholate was selected as it interacts with AmB forming micelles, while it accumulates in buccal tissue after penetration without causing a loss of superficial cell layers and interacts with the intercellular or membrane lipids increasing the permeability of drugs through the epithelium [10]. Design of experiment (DoE) studies enabled us to identify the optimal drug:excipient ratio needed to ensure high drug loading (1% AmB) that is critical in achieving local concentrations well above the $IC_{50}$ against *Candida albicans*. The optimised ODF were fully characterised and their *in vitro* antifungal activity evaluated.

## 2. Materials and Methods

### 2.1. Materials

AmB was purchased from Azelis (Barcelona, Spain). Hydroxypropylmethyl cellulose acetate succinate (HPMC AS 912, Affinisol™), maltodextrin (Glucidex 12D) and sorbitol were a gift from DowPharma (Dewsbury, UK) and Roquette (Valencia, Spain). Microcrystalline cellulose (Avicel 200) and microcrystalline cellulose and sodium carboxymethylcellulose (Avicel CL-611) were kindly donated by FMC (Cork, Ireland), while hydroxypropyl cellulose (HPC, Klucel HXF) was a gift from Ashland (Barcelona, Spain). Dextrose and dextran from Leuconostoc mesenteroides (16 KDa) were purchased from Sigma (Madrid, Spain). Humidity capsules and stability chambers were purchased from Amebis Limited (Dunshaughlin, Ireland). All other chemicals and solvents were at least of ACS reagent grade and were used without further purification.

## 2.2. *Quality by Design (QbD) Optimisation of ODF*

Several critical quality attributes (CQAs) such as disintegration time in artificial saliva and the physical characteristics of the ODF (burst strength, flexibility, tackiness) were identified as key factors in order to meet the Target Product Profile (TPP) (Tables S1 and S2 in Supplementary Materials). A Taguchi design (L8 = 2^7) was carried out using Design Expert software 8.04 (Stat-Ease, Minneapolis, MN, USA). Seven formulation variables (factors) and two levels of each factor affecting the film formation were investigated (Table 1). Disintegration time, burst strength and appearance were evaluated as responses (Table 2).

**Table 1.** Formulation and process variables with their respective high and low levels included in the Taguchi screening design.

| Variables | Levels | |
|---|---|---|
| | **Low (−1)** | **High (+1)** |
| Type of dextrose-derived-polymer film former | Maltodextrin | Dextran |
| Taste masking agent (5%) | Sorbitol | Dextrose |
| Type of binder/suspending agent | Avicel 200 | Avicel CL-611 |
| Amount of Avicel polymers (%) | 0 | 25 |
| Amount of total plasticisers (PEG 400:glycerol, 1:1 *w/w*) (%) | 5 | 20 |
| Methanol | Low (1–2 mL) | High (>3 mL) |
| Amount of cellulose-derived film formers (HPMC 912 and HPC in a 1:4 weight ratio) (%) | 0 | 15 |

### 2.2.1. ODF Manufacture and Response Evaluation

Eight formulations of AmB-loaded ODFs (3 g each) (Table 2) were prepared as follows: dextrose-derived-polymer film former, Avicel 200 or Avicel CL-611, taste masking agent (5%) and plasticisers were weighed and mixed in a mortar and pestle. To this mixture, 3 mL of freshly prepared AmB-loaded micelles (30 mg of AmB and 24.6 mg of sodium deoxycholate [11]) was added and manually mixed until a homogenous mixture was formed. The cellulose-derived film formers were then added, if required, and mixed. Methanol was added to reduce the viscosity of the mixture to a pourable homogenous suspension that was immediately cast onto a release liner (Primeliner 36 μm 1S, Loparex BV, Apeldoorn, The Netherlands) using a coating knife (Multicator 411, Erichsen, Hemer, Germany) and film applicator (Erichsen Coatmaster 510 film applicator, Erichsen, Hemer, Germany), at a speed of 10 mm $s^{-1}$ under vacuum (air pressure 60 Pa), to form a wet film with a thickness of 1000 μm. The film was allowed to dry under vacuum for 4–5 h. Once the films were dried, they were carefully removed from the release liner and properties of the film were evaluated (Table 2).

**Table 2.** Design matrix of AmB-loaded orodispersable films (ODFs) prepared as per the Taguchi design. The overall appearance of the film was evaluated as the sum of the folding endurance, adhesion and homogeneity. Regarding the folding endurance, a value of 1 was assigned for films with good/flexibility able to fold 180 degrees without breaking above 50 folds, 2 as medium flexibility films (25–50 times) and 3 as poor flexibility films, which break easily when folded more than 25 times. Adhesion of the dried film to the release liner was also taken into account, assigning a number from 1 to 3—with 1 allowing intact removal of the film and 3 complete breakage of the film upon removal. The homogeneity was ranked taking into consideration the value of 1 for films that exhibit a smooth surface with no cracks, 2 for films with some irregular surfaces and 3 for non-homogeneous films with lumps and/or cracks on the surface.

| Run | Type of Dextrose-Derived- Polymer Film Former | Taste Masking Excipient | Type of Avicel | Amount of Avicel (%) | Amount of Total Plasticisers (%) | Amount of Methanol | Amount of Cellulose-Derived Film Formers (%) | Disintegration Time (Seconds) | Flexibi-lity | Adhesion | Appea-rance | Burst Strength (Mn mm) |
|---|---|---|---|---|---|---|---|---|---|---|---|---|
| **1** | Dextran | Sorbitol | 200 | 25 | 5 | 1 | 15 | 270 | 2 | 1 | 4 | 1300.9 |
| **2** | Maltodextrin | Dextrose | 200 | 25 | 20 | 2 | 15 | 90 | 1 | 1 | 3 | 2468 |
| **3** | Maltodextrin | Dextrose | 200 | 0 | 5 | 1 | 0 | 60 | 2 | 3 | 6 | 0 |
| **4** | Dextran | Dextrose | CL611 | 25 | 5 | 2 | 0 | 360 | 3 | 2 | 8 | 0 |
| **5** | Maltodextrin | Sorbitol | CL611 | 25 | 20 | 1 | 0 | 165 | 2 | 3 | 6 | 0 |
| **6** | Maltodextrin | Sorbitol | CL611 | 0 | 5 | 2 | 15 | 150 | 2 | 1 | 4 | 1061 |
| **7** | Dextran | Sorbitol | 200 | 0 | 20 | 2 | 0 | 15 | 3 | 1 | 7 | 0 |
| **8** | Dextran | Dextrose | CL611 | 0 | 5 | 1 | 15 | 240 | 3 | 2 | 8 | 0 |

Disintegration times of ODFs (1 × 1 cm) were measured in 3 mL of artificial saliva prepared as previously described: 14.4 mM sodium chloride, 16.1 mM mg potassium chloride, 1.31 mM calcium chloride dihydrate, 0.54 mM magnesium chloride hexahydrate, 1.96 mM dibasic potassium phosphate adjusted to pH 5.7 ± 0.01 [12] under gently shaking (30 slow 90° inversions of the vial per min). The mechanical properties (burst strength) of ODFs were evaluated using a texture analyser (Texture Analyser TA-XTplus, Stable Microsystems, Godalming, UK) attached to a film support rig (HDP/FSR, Stable Microsystems) [13]. For burst strength, the force required to rupture or break films was measured using a 5 mm spherical stainless-steel ball probe with probe adapter which was connected to the load cell. A film (35 × 15 mm) was placed in a film supporting rig and the moving probe reached the surface of the film with a pre-test speed of 2 mm $s^{-1}$, test speed of 1 mm $s^{-1}$ and post-test speed of 10 mm $s^{-1}$. The force applied had a trigger load of 4.9 N and the force maximum (mN), travel distance (mm) and area under the curve (mN mm) were measured. Finally, the overall appearance of the films was evaluated as the sum of folding endurance, adhesion of the dried film to the release liner and homogeneity. Folding endurance was manually measured by counting the number of times the film could be folded at a 180° angle to the plane without breaking. A value of 1 was assigned for films with good/flexibility able to fold 180 degrees without breaking above 50 folds, 2 as medium flexibility films (25–50 times) and 3 as poor flexibility films, which break easily when folded more than 25 times. Adhesion of the dried film to the release liner was also taken into account, assigning a number from 1 to 3, with 1 indicating intact removal of the film and 3 complete breakage of the film upon removal. The homogeneity was also visually inspected, assigning 1 for films that exhibit a smooth surface with no cracks, 2 for films with some irregular surfaces and 3 for non-homogeneous films with lumps and/or cracks on the surface.

#### 2.2.2. ODF Optimisation

Mathematical modelling was carried out by multiple linear regression analysis (MLRA). Only the statistically significant coefficients ($p < 0.05$) were considered in framing the polynomial equations, and the model was evaluated by analysing the $p$-value, coefficient of correlation ($R^2$) and predicted residual sum of squares (PRESS) [14]. Films were optimised in order to minimise the disintegration time and improve overall appearance and mechanical strength.

### 2.3. *Full Physicochemical Characterisation of Optimised ODF*

A full evaluation was performed including particle size and zeta potential (after disintegration in artificial saliva), powder X-ray Diffraction (PXRD), Fourier Transformed Infrared (FT-IR), Dynamic Vapour Sorption (DVS), Scanning Electron Microscopy (SEM) and surface area [11,14,15]. Briefly, particle size and zeta potential were measured in a Zetatrac after dilution (1 to 100) with artificial saliva. XRD measurements ($n = 3$) from 5° to 40° (2 theta) and a step scan rate 0.05° per second were performed in a Miniflex II Rigaku diffractometer with Ni-filtered Cu K$\alpha$ radiation (1.54 Å) using a tube voltage and tube current of 30 kV and 25 mA respectively. FT-IR spectra were scanned in the range of 650–4000 $cm^{-1}$ with a resolution of 4 $cm^{-1}$ on a PerkinElmer Spectrum 1 FT-IR Spectrometer equipped with a UATR and a diamond/ZnSe crystal accessory. Baseline correction and data normalization were performed using Spekwin32 version 1.71.6.1. Water sorption kinetic profiles were obtained using a DVS (Advantage, Surface Measurement Systems, Alperton, UK) at 25.0 ± 0.1 °C. Samples (10–20 mg) were dried at 0% relative humidity (RH) for 1 h followed by step changes of 10% RH up to 90% RH, and the reverse for desorption. SEM was carried out in a Zeiss Supra Variable Pressure Field Emission Scanning Electron Microscope (Oberkochen, Germany) equipped with a secondary electron detector at 15 kV. Surface area was determined by the Brunauer, Emmett, Teller (BET) isotherm method using N2 adsorption with 6 points in the relative pressure range of 0.05–0.3 in a Micromeritics Gemini VI surface area analyser (Particular Sciences Ltd., Dublin, Ireland).

*2.4. Content Uniformity*

ODF (1 × 1 cm) were weighed and disintegrated in 3 mL of deionised water prior to being diluted (1 to 2) with methanol to ensure full drug solubilisation. Samples were centrifuged to precipitate undissolved excipients (5000 rpm for 10 min) and AmB dissolved in the supernatant was quantified using a validated HPLC method [16]. Experiments were performed using films from ten different sections of the cast ODF (15 × 10 cm).

*2.5. Release Studies in Artificial Saliva*

ODF (1 × 1 cm) were dissolved in 10 mL of artificial saliva at 37 °C under slow magnetic stirring (50 rpm). Samples (1 mL) were obtained at 1, 2, 4, 6, 8, 10, 15 and 30 min and media were replaced with fresh artificial saliva. Withdrawn samples (0.1 mL) were mixed with methanol (0.1 mL), vortex and centrifuged (5000 rpm, 10 min) for HPLC analysis to determine the percentage of drug released. The remaining sample volume was centrifuged (500 rpm, 1 min) to precipitate undissolved excipients such as microcrystalline cellulose. The particle size of the supernatant was measured using a Malvern Zetasizer (Malvern Nano Zs, Malvern Instruments, Malvern, UK). Additionally, the supernatant was analysed by UV (300–450 nm, Multiskan GO, Thermo Scientific, Basingstoke, UK) to determine the aggregation state of the AmB [17]. The ratio of the absorbance at 332 nm corresponding to dimeric AmB versus the absorbance at 408 nm that corresponds to the monomeric state was plotted to demonstrate the prevalence of the dimeric aggregation state of AmB.

*2.6. In Vitro Antifungal Assays*

In vitro antifungal activity was tested based on the agar diffusion assay as described by Ruiz et al. [1] according to the National Committee for Clinical Laboratory Standards (NCCLS) Method for antifungal disk diffusion susceptibility testing of yeast, standard M44-A2 [18]. In vitro activity was tested on three different *Candida* spp. (*C. albicans* CECT 1394, *C. parapsilosis* 57744 and *C. krusei* 52009 which was kindly provided by Dr. Pérez (CAQYM, University of Alcala de Henares, Alcalá de Henares, Spain) [1]. Strains were cultured in Sabouraud dextrose agar for 72 h to ensure viability and absence of contamination at 35 °C (±2 °C). Antifungal tests were carried out in Müeller Hinton agar (MHA) supplemented with glucose (2% *w/v*) and methylene blue (0.5 mg/mL). Inoculum was prepared by picking a few distinct colonies, which were suspended in 3 mL of sterile saline (0.9%). The resulting suspension was vortexed and its turbidity was adjusted with a spectrophotometer by adding sufficient sterile saline or more colonies to adjust the transmittance to that produced by a 0.5 McFarland standard at 530 nm wavelength, resulting in a yeast stock suspension of $1 \times 10^6$ cells per mL. Yeast suspension was inoculated to the MHA (200 mL) and was casted in disposable sterile petri dishes (instead of spreading it on the surface of the plate as specified in M44-A2). Once solidified, AmB ODFs (circles with a 6 mm diameter) were tested. Four disks were placed in each plate. AmB ODF in vitro activity was compared to commercially available AmB Neo-Sensitabs tablets (10 μg, 6 mm tablets from Rosco diagnostic A/S, Taastrup, Denmark) and AmB impregnated on inoculation 6 mm paper disks (10 μg/20 μL of DMSO) with appropriate DMSO controls. Once the disks were placed on the surface of the agar, the plates were inverted and placed in an incubator set to 35 °C (±2 °C) within 15 min after the disks were applied. After 24 h of incubation, the inhibition halo was measured.

*2.7. Stability Studies*

Physicochemical stability studies were performed under accelerated conditions (40, 60 and 80 °C) for one week and long term at 5 ± 3 °C and 25 ± 3 °C. AmB ODF (1 × 1 cm) were placed in sealed vials into Amebis chambers (Amebis Ltd., Dublin, Ireland) at the selected temperature. A sensor cap was used to seal the test chamber and a logger cap connected to the sensor cap was used to collect and transmit the temperature and humidity test conditions wirelessly to the Amebis Control Software [14]. Disintegration time and drug content were quantified at different time points (time zero, day 1, 3 and

7). The degradation rate of AmB was calculated by fitting the percentage of drug degraded at different time points to several degradation kinetic equations (zero order, first order, second order, Avrami and diffusion) and the best fitted degradation kinetic model was selected (i.e., highest $R^2$). Using the degradation rates at different temperatures, the Arrhenius equation was employed to calculate the activation energy (Equation (1)):

$$K = A\, e^{\frac{-Ea}{RT}} \quad (1)$$

where $K$ is the degradation rate (% drug degraded/day), $A$ is the collision factor, $T$ is the absolute temperature in Kelvin, $R$ is the gas constant (1.985 cal/mol/K) and $Ea$ is the activation energy in cal/mol [14]. Drug stability at room temperature was then predicted using the Arrhenius equation and compared to experimental values.

## 3. Results

### 3.1. QbD Studies for Optimisation of AmB-Loaded ODFs

The first-order mathematical model generated for each response variable was found to be statistically significant ($p < 0.05$ in each case). Co-efficients with $p$ values > 0.1 were considered insignificant based on Pareto charts and ANOVA analysis. High $R^2$ values for the polynomial equations obtained for all the response variables indicate a good fit to experimental data (Table S3). The variables, type and amount of Avicel and number of plasticizers, showed a significant effect on the disintegration time of the ODFs (Figures S1 and S2). ODFs with lower disintegration time were obtained when lower amounts of Avicel and cellulose-derived film formers were used. The disintegration time was reduced when higher percentages of plasticisers were employed and when Avicel 200 was incorporated in the film compared to Avicel CL-611, as the latter acts as a viscosity enhancing agent resulting in thixotropic gels, which retard disintegration (Figure 1A,B).

The choice of dextrose-derived film former, volume of methanol added, type/amount of Avicel and amount of cellulose-derived film formers had a significant impact on burst strength (Figures S3 and S4). The use of maltodextrin and Avicel 200 resulted in films with higher burst strength. The higher the amount of Avicel, HPMC AS/HPC and methanol, the better the mechanical strength of the films (Figure 1C,D).

The appearance of the films was rated as described above (Figure S5). DoE indicated that the main variables that contributed to the appearance of the film were: the type of dextrose-derived film former and the amount of cellulose-derived film formers utilised, followed by type and amount of Avicel and taste masking agent (Figures S6 and S7), although results were not statistically significantly different. The appearance of the film was smoother when maltodextrin was used, although high amounts of maltodextrin increased film tackiness. When higher amounts of Avicel and cellulose-derived film formers were used, appearance was improved. Avicel 200 and sorbitol resulted in smoother film surfaces compared to dextrose films as dextrose can recrystallize resulting in rougher surfaces [19] (Figure 1E,F).

### 3.2. Manufacturing of Optimised AmB ODF

A trade-off between key CQAs was necessary to attain optimal characteristics, i.e., short disintegration time (which is critical for faster onset of action), maximal burst strength (in order to obtain robust films that can be easily manufactured and packaged without breaking), and good appearance. A closer match to ideal CQAs was obtained with sorbitol, Avicel 200 (10%), 20% of plasticisers (PEG 400:glycerol, 1:1 *w/w*) and 10% of cellulose-derived film formers (HMPC 912 AS:HPC, 1:4 weight ratio). Regarding the type of dextrose-derived film former, maltodextrin conferred better flexibility and faster release from the films compared to those obtained with dextran, as dextran interacts more strongly compared to microcrystalline cellulose or the modified cellulose-derived film formers. However, high amounts of maltodextrin significantly increased adhesion of the films to the release liners. Thus, we decided to optimise the ODF using a mixture of dextran and maltrodextrin (1:1

*w/w*) (Table 3). The films exhibited a good overall appearance with a dried thickness of 0.14 ± 0.01 mm, a weight of 28.5 ± 1.5 mg/cm$^2$, a drug content of 0.996 ± 0.045 mg/g (which was uniform with a low standard deviation and a variance coefficient of 4.5%), a burst strength of 2190 mN mm and a disintegration time of ≤60 s in 3 mL of artificial saliva.

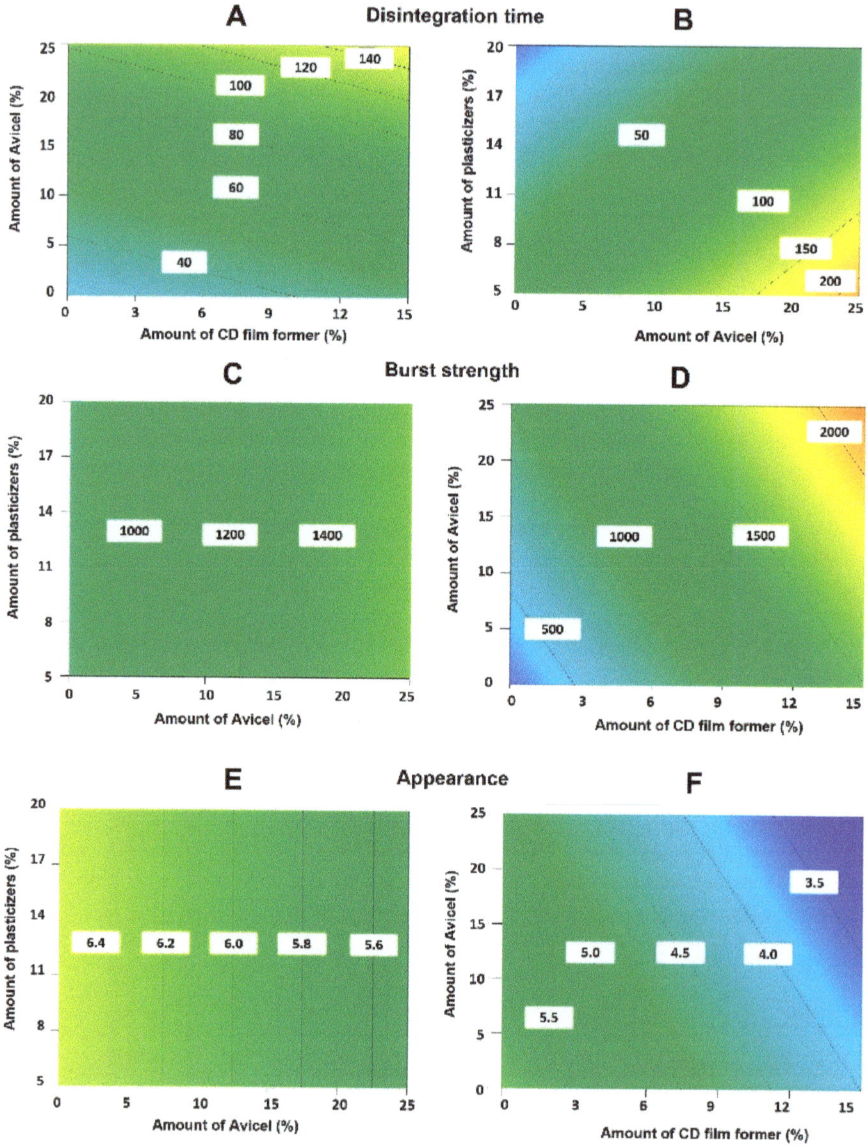


**Figure 1.** Contour plots showing the influence of the most influential factors affecting the disintegration time (**A**,**B**), the burst strength (**C**,**D**) and the overall appearance (**E**,**F**) of the AmB-loaded orodispersable films (ODFs). Key: CD, cellulose-derived.

**Table 3.** Composition and properties of the optimised AmB-loaded ODFs.

| Composition | |
|---|---|
| Dextran | 25% (0.74 g) |
| Maltodextrin | 25% (0.74 g) |
| Sorbitol | 5% (0.15 g) |
| Avicel 200 | 10% (0.3 g) |
| Plasticisers (PEG 400:Glycerol, 1:1 *w/w*) | 20% (0.3 g + 0.3 g) |
| Simil Fungizone (9) | 3 mL (=30 mg AmB + 24.6 mg sodium deoxycholate) |
| Cellulose-derived film formers | 15% (0.09 g HMPC AS + 0.36 g HPC) |
| Methanol | 1.5 mL |
| **Properties** | |
| Disintegration time | 60 ± 3 s |
| Burst strength | 2190 ± 140 mN mm |
| Overall score of appearance | 4 = Flexibility (1), tackiness (2) and homogeneity (1) |

SEM micrographs revealed a smooth and porous surface of the optimised AmB-loaded ODF compared to films obtained in the DoE experiment 1 and 2, which exhibited a granular rough texture or embedded crystals respectively probably due to the use of dextran in the first film instead of maltodextrin and the presence of dextrose in DoE 2 ODF which tends to crystallise (Figure 2). Films obtained in DoE 6 experiments appeared cracked and exhibited low elasticity likely due to the low percentage of plasticisers and Avicel 200 included in the formulation.

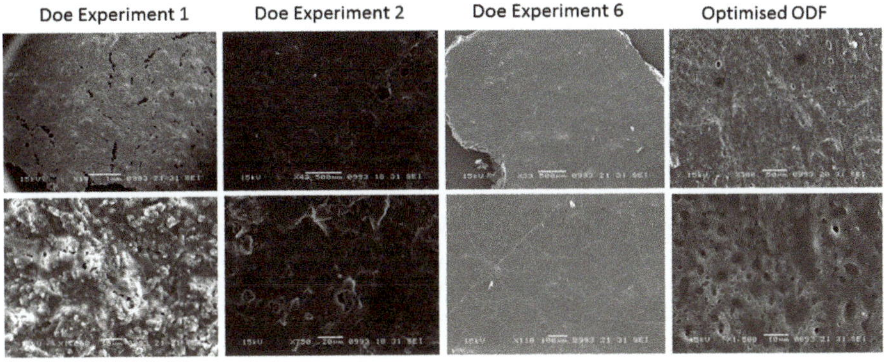


**Figure 2.** Scanning Electron Microscopy (SEM) micrographs of optimised AmB-loaded ODFs compared to those obtained in the Taguchi DoE before optimization. Bars; DOE Experiment 1—Top: 1 mm and Bottom: 10 µm, DOE Experiment 2—Top: 500 µm and Bottom: 20 µm, DOE Experiment 6—Top: 500 µm and Bottom: 100 µm, Optimised ODF—Top: 50 µm and Bottom: 10 µm.

## 3.3. *Performance and Further Characterisation of the Optimised AmB-Loaded ODF*

The smooth and porous surface observed by SEM can be associated with the high 2.3 ± 0.5 $m^2/g$ surface area of the optimized film (Figure 2). AmB-loaded ODFs showed a 40% increase in mass at 90% relative humidity, associated with a large water uptake due to the films hydrophilicity as shown by the water sorption kinetic profile (Figure 3A). No mass loss (associated to phase transformation/crystallization) was observed during the sorption or desorption cycle. The solid state of the film was retained (as demonstrated by post DVS XRD analysis (Figure S8) indicating an overall acceptable physical stability. The FTIR spectra showed a broadening of the peak at 1691 $cm^{-1}$ (C=O stretch) probably attributed to hydrogen bonding between the AmB and excipients such as sodium deoxycholate, sorbitol and acetate succinate groups of the HPMC AS (Figure 3B and Figure S9).

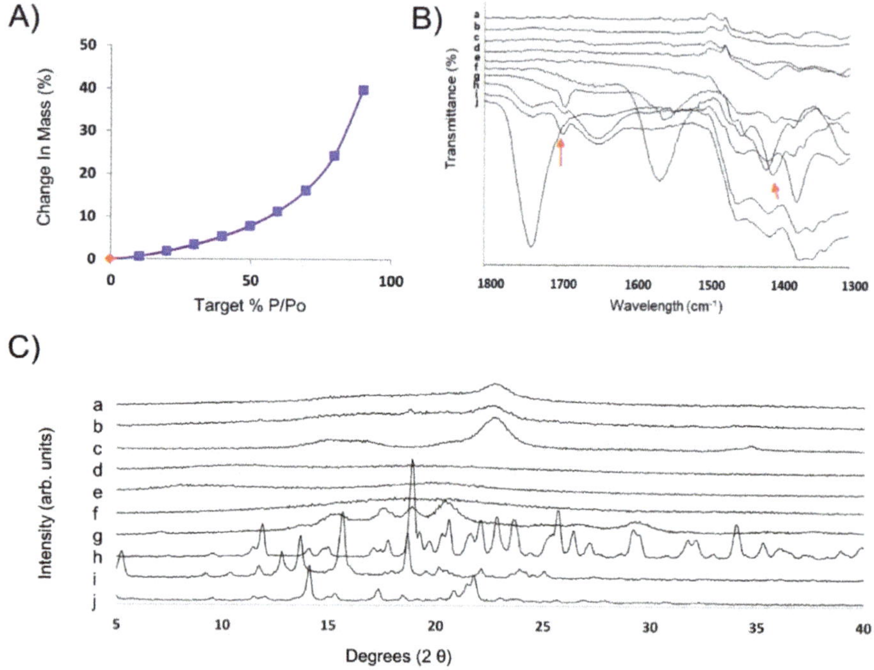


**Figure 3.** Physicochemical characterization of optimised AmB-loaded ODFs. (**A**) Water sorption kinetic profile. (**B**) FTIR spectra: (a) hydroxypropyl cellulose (HPC), (b) maltodextrin, (c) dextran, (d) Avicel 200, (e) sorbitol, (f) sodium deoxycholate, (g) AmB, (h) AmB-loaded ODF, (i) physical mixture, (j) HPMC 912 AS. (**C**) PXRD patterns: (a) AmB-loaded ODF, (b) Physical mixture of all components, (c) Avicel 200, (d) HPMC 912 AS, (e) HPC, (f) maltodextrin, (g) dextran, (h) sorbitol, (i) sodium deoxycholate, (j) AmB.

The XRD pattern of the physical mixture of all components revealed crystalline Bragg peaks attributed to sorbitol (Figure 3B(b,h)), which are not present in the casted film. However, a characteristic halo attributed to the semi-crystalline nature of the microcrystalline cellulose (Avicel 200) was observed in the optimised ODFs. Lower intensity values at several Bragg peaks (5°, 14.15°, 17.35°, 21.75°) were observed in the diffractogram of the AmB-loaded ODF compared to the physical mixture, which can be attributed to the presence of amorphous AmB-sodium deoxycholate complexes in the films. However, bearing in mind that the AmB content in the ODF is 1%, it is likely that the results obtained from XRD measurements are not conclusive due to the XRD detection limit for crystalline AmB.

### *3.4. Release and Aggregation State*

Aligned with the high porosity of the films, ODF presented a fast-dissolving behaviour (>80% in 10 min) in saliva (Figure 4A). Once disintegrated, particle size and zeta potential of the resulting suspension was measured. Initial particle size was bimodal mainly dominated (>70%) by large particles (>1 µm) due to insoluble excipients. After centrifugation, a white pellet and a transparent yellow supernatant was obtained. The latter demonstrated a bimodal particle size distribution of 8.7 ± 2.5 nm and 918 ± 120 nm with an anionic zeta potential of −14 ± 3 mV, indicating that after disintegration of the ODFs, AmB remained solubilised in the supernatant in equilibrium between micelles and particles close to 1 µm in size. TEM images confirmed the presence of micelles and particles of that size (Figure 4(B1)). Characteristic crystals corresponding to unprocessed AmB (Figure 4(B2)) were not observed after disintegration of the ODF and release of the AmB in the media which supports drug solubilisation within micelles.

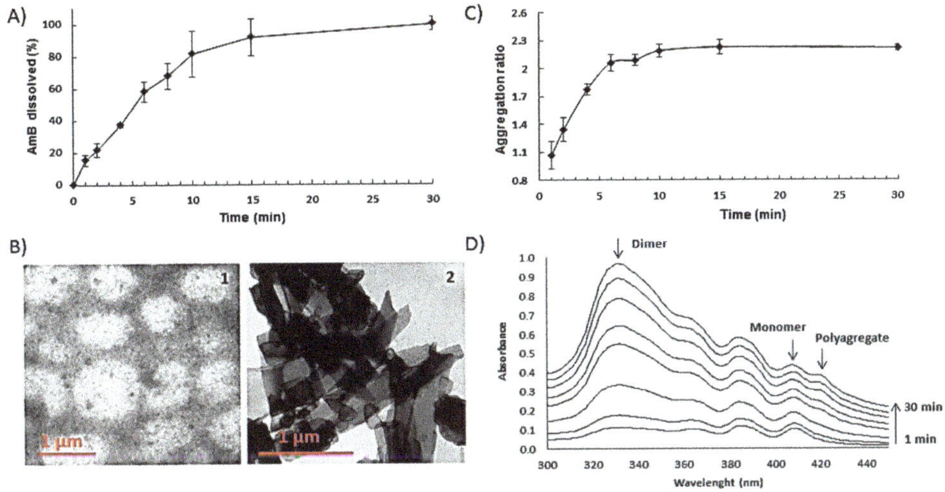


**Figure 4.** Release profile, morphology and aggregation state of optimised AmB loaded ODFs. Key: (**A**) Release profile of AmB-loaded ODFS in artificial saliva. (**B**) Transmission electron microscopy (TEM) of AmB-loaded ODF after reconstitution in aqueous media (left image) and crystalline unprocessed AmB (right image); (**C**) Aggregation ratio during the release studies; (**D**) Transformation of AmB aggregation states over time.

Regarding the aggregation state of AmB released from the films, UV spectroscopy indicated a shift from the monomeric to dimeric state illustrated by the faster increase in the peak at 332 nm compared to the one at 408 nm corresponding to the monomeric state (Figure 4C,D). At the earlier time points (1–2 min), there is an equilibrium in solution between monomer and dimer. As time progresses and a higher percentage of AmB is released to the media, AmB self-aggregates shifting from monomer to dimer after 2 min and from dimer to polyaggregate after 8 min (peak at 420 nm) [17,20]. AmB solubility can be enhanced by the presence of PEG 400 and sodium deoxycholate in the formulation [21]. Fungizone®, a commercial AmB parenteral formulation, consists of 1:2 AmB:sodium deoxycholate molar fraction that facilitates the solubilisation of AmB in dimeric form within micelles [20]. The presence of nanometric (≈10 nm) spherical single layer particles in solution after ODF disintegration can be explained by the partial solubilisation of the drug within sodium deoxycholate micelles. However, even though initially AmB is incorporated in the solid mixture of excipients solubilised in sodium deoxycholate, the later addition of solvent to produce a pourable castable formulation could destabilise initially formed AmB sodium deoxycholate micelles. Thus, the different aggregation states of released AmB can be a consequence of the initial solubilisation of AmB in the monomer state in the presence of PEG 400 and the formation of ion pairs with sodium deoxycholate, followed by the encapsulation of AmB within micelles resulting in prevalence of the dimeric state. When higher amounts of AmB are released over time in the media, AmB self-aggregates into polyaggregates close to 1 µm in size which are stabilised by the release of dextrose-based polymers, in particular maltodextrin, which has been previously shown to be able to effectively stabilise dispersed systems such as oil-in-water emulsions [22].

### 3.5. Antifungal Activity

The optimised AmB-loaded ODFs showed good in vitro antifungal activity against the three *Candida* spp. (with an inhibition zone >15 mm) equivalent to that of AmB dissolved in DMSO and the commercially available disks, which indicates good drug release from the film and diffusion across the agar of the solubilised AmB (Figure 5). Bearing in mind the potency of the drug against *C. albicans*, *C. parapsilopsis* and *C. krusei* ($MIC_{50}$ ranges from 0.25–1 µg/mL) reported in the literature [2], the dose delivered by a 1 × 1 cm film would be adequate to ensure efficacy against buccal candidiasis. Considering that the drug loading is 1% *w/w* and that a 1 $cm^2$ film weights around 30 mg, each 1 $cm^2$ film contains approximately 0.3 mg of AmB. Assuming a volume of ~1 mL in the oral cavity [3,23], the concentration of AmB would be 300 μg/mL, which is well above the $MIC_{50}$ reported in literature. Even if the concentration is diluted further with 10 mL or 100 mL i.e., 10 or 100-fold, the AmB concentration would be in range between 3 or 30 μg/mL, which are concentrations still above the $MIC_{50}$ and, for this reason, we believe that the drug levels would be adequate to elicit a pharmacological effect.

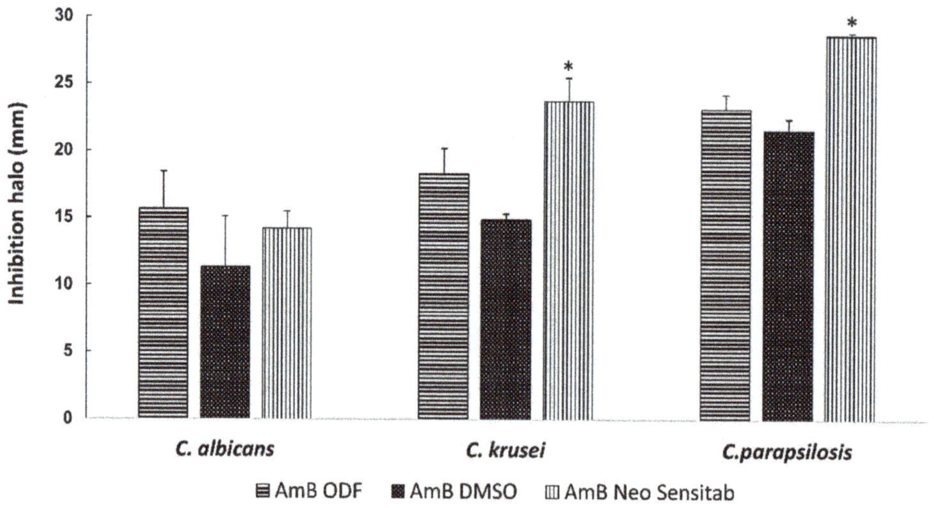


**Figure 5.** In vitro activity against *C. albicans*, *C. krusei* and *C. parapsilopsis*. The isolates were classified as susceptible (S) to AmB when the inhibition zone was ≥15 mm, resistant (R) when it was ≤10 mm and intermediate (I) or susceptible-dose dependent when the inhibition zone was between 10- and 15-mm. Inhibition zone diameters are expressed as mean ± SD in mm. All experiments were performed in triplicate. * $p < 0.05$ One-way ANOVA test.

### 3.6. Physicochemical Stability

Stability studies showed that the optimised AmB-loaded ODF was physiochemically stable over a year (>90% drug content) at 5 °C under desiccated conditions. Accelerated stability studies demonstrated a pronounced chemical degradation (80 °C >> 60 °C >> 40 °C), while a change in colour occurred at the highest temperature (Figure 6). The disintegration times of ODFs subject to accelerated stability studies was reduced over time, probably due to the evaporation of bound water within the film and the formation of micropores (Figure 6). However, in all the tested conditions, the disintegration time remained below 2 min as considered appropriate for fast-disintegrating ODFs. The Avrami kinetic model fitted the degradation of AmB from loaded ODFs. This kinetic model is commonly applied to evaluate the growth and formation of crystals [24]. Nevertheless, several authors have also used this model in the stability prediction of nanocomposites [25,26]. The breaking of a 3D network becomes more heterogeneous as the degradation progresses, and thus, the degradation proceeds faster as the combined effects of physical and chemical degradation occur simultaneously, especially at higher temperatures [26]. The activation energy of the AmB-loaded ODFs was 5.71 Kcal/mol, which correlates with the poor physicochemical drug stability observed at high temperatures (Figure 6). Experimental and predicted data from the Arrhenius equation showed that ODFs stored at 25 °C at desiccated conditions remained stable over 15 days (>90% drug content, disintegration time of 55 s). Thus, our AmB ODFs would remain stable for 2 weeks at room temperature in a closed pouch, which is a significant advantage to existing AmB commercialised parenteral formulations that are unstable after 24 h of reconstitution of the lyophilised powder [4] reducing overall cost of treatment and wastage.

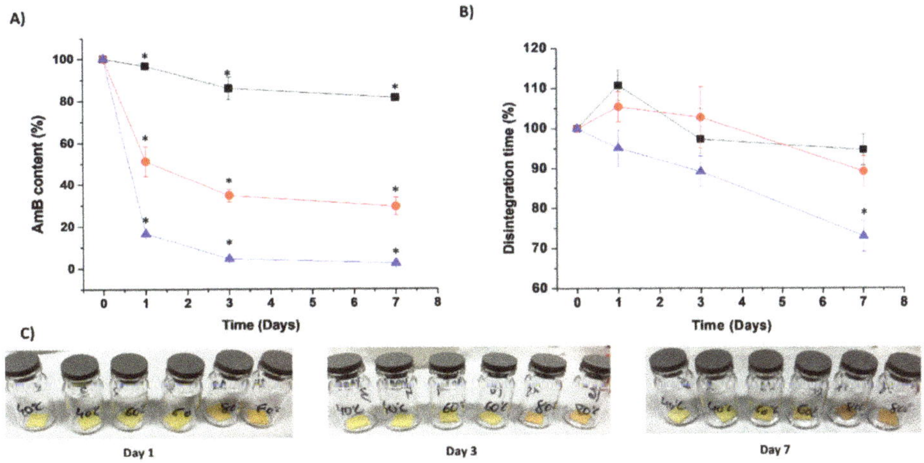


**Figure 6.** Physicochemical stability studies of optimised AmB-loaded ODFs at different temperatures. Key: -■- 40 degrees, -●- 60 degrees, -▲- 80 degrees. (**A**) Chemical stability expressed as AmB content. (**B**) Change in disintegration time for films stored at different temperatures. (**C**) Physical appearance. * $p < 0.05$ indicative of statistically significant difference compared to AmB content and disintegration time of films at time 0 (One-way ANOVA with a post-hoc Tukey's test, level of significance set at 5%).

## 4. Discussion

To the best of our knowledge, this is the first report of AmB, a high molecular weight poorly water-soluble antifungal, being formulated as an ODF for the treatment of oropharyngeal candidiasis. The optimised ODF was achieved by combining dextran and maltodrextrin as dextrose-derived polymer film formers with sorbitol for taste masking, microcrystalline cellulose (Avicel 200) for enhancing mechanical strength, PEG 400 and glycerol as plasticizers to ensure faster disintegration and adequate plasticity and HMPC 912 AS /HPC for facilitating film formation. The combination of PEG 400 and glycerol (1:1 weight ratio) was selected based on previous results (data not shown). Glycerol improves mouthfeel, which is a necessity in an ODF formulation, as well as enhancing the viscosity resulting in higher, unacceptable disintegration times. In contrast, PEG 400 decreases disintegration time of ODF and, in combination with HPMC, gives optimal films in terms of tensile strength and flexibility [27].

Currently, nystatin (which has a similar chemical structure to AmB) formulated as a suspension is considered as the reference treatment for oral candidiasis; nevertheless, a recent meta-analysis demonstrated a limited efficacy for nystatin, lower than fluconazole, in treating oral candidiasis in infants, children and HIV patients, which is related to a poor and variable bioavailability in the oral mucosa [28]. AmB has shown 4-fold higher in vitro efficacy than nystatin [29], but its poor aqueous solubility limits its use in clinical practice, and it is only marketed as intravenous formulations. After disintegration of the ODF, AmB is maintained solubilised in the aqueous media in equilibrium between different aggregation states mainly monomer and dimer at earlier times points, which have shown to possess much greater efficacy than polyaggregates against *Candida* [30,31]. Unlike previous studies in which the antifungal activity was tested after disintegration of the films in liquid form [32], enhanced in vitro activity of the AmB-loaded ODF was shown here for the intact film placed on top of the agar, showing that the release and diffusion of the drug across the agar occurs and hence AmB is free to elicit its effect. Excipients utilised in ODFs are used within ranges that are considered GRAS (generally regarded as safe). Sodium deoxycholate, at the concentrations used in optimised the ODFs, has been shown to cause no major morphological changes [10]. Although further toxicological and in vivo efficacy studies are required, we have shown that this technology platform can be a promising AmB formulation for the treatment of oral candidiasis.

## 5. Conclusions

Fast-dissolving orodispersible films containing a very poorly soluble and unstable drug, AmB, have been successfully engineered and prepared using a solvent casting film method in which the drug is solubilised at a final concentration of 1% *w/w* upon disintegration of the film. The optimised AmB orodispersible flms consisted of a mixture of dextrose-derived polymers (25% dextran and 25% maltodextrin) and cellulose-derived film formers (3% HMPC AS and 12% HPC) with 10% Avicel 200 added to enhance the mechanical strength of the film, 5% sorbitol for taste masking, and 10% PEG 400 and 10% glycerol as plasticizers. The optimised ODF exhibited a fast disintegration time (60 ± 3 s), quick release in artificial saliva (>80% in 10 min), high burst strength (2190 mN mm), good chemical stability (1 year at refrigerated conditions and 2 weeks at room temperature) and high efficacy against several *Candida* spp. (*C. albicans*, *C. parapsilosis* and *C. krusei* with an inhibition halo >15 mm). Although further toxicological and in vivo efficacy studies are required, these novel orodispersible films prepared with GRAS excipients have potential in the treatment of oro-and pharyngeal candidiasis, and hence represent a promising system with wide applications in clinical practice among immunocompromised patients suffering from this disease.

**Supplementary Materials:** The following are available online at http://www.mdpi.com/1999-4923/11/8/369/s1, Figure S1: Pareto charts depicting the effect of (C) Type of Avicel, (D) Amount of Avicel, (E) Amount of plasticisers and (G) Amount of cellulose-derived film formers on the disintegration time. Orange colour indicates a positive effect whereas blue colour indicates a negative effect. Figure S2: Effect of the four variables (Type of Avicel, amount of Avicel, amount of plasticizers and amount of cellulose-derived film formers) on the disintegration time. Figure S3: Pareto charts depicting the effect of (A) Type of dextrose-derived film former, (C) Type of Avicel, (D) Amount of Avicel, (F) Volume of methanol and (G) Amount of cellulose-derived film formers on the burst strength of the film (expressed as AUC). Orange colour indicates a positive effect whereas blue colour indicates a negative effect. Figure S4: Effect of the significant variables (Film former, Type of Avicel, amount of Avicel and amount of cellulose-derived film formers) on the burst strength expressed as AUC of the film. Figure S5: Appearance of the eight AmB-loaded films prepared according to Taguchi matrix design. Figure S6: Pareto charts depicting the effect of (A) Type of dextrose-derived film former, (B) Taste masking, (C) Type of Avicel, (D) Amount of Avicel and (G) Amount of cellulose-derived film formers on the appearance of the film. Orange colour indicates a positive effect whereas blue colour indicates a negative effect. Figure S7: Effect of the five variables with higher impact on the final appearance of the film (Type of Avicel, amount of Avicel, taste masking agent, type of dextrose-derived film former and amount of cellulose derived-film formers). Figure S8: PXRD patterns of raw materials and AmB-loaded ODF before and after DVS analyses. Key: AmB ODF post DVS, (b) AmB ODF, (c) physical mixture, (d) Avicel 200, (e) HPMC AS; (f) HPC, (g) maltodextrin, (h) dextran, (i) sorbitol, (j) sodium deoxycholate, (k) AmB. Figure S9: FT-IR spectra. (a) HPC, (b) maltodextrin, (c) dextran, (d) Avicel 200, (e) sodium deoxycholate, (f) sorbitol, (g) HPMC 912 AS, (h) AmB, (i) physical mixture, (j) AmB-loaded. Table S1: Target product profile (TPP) elements for AmB-loaded ODFs. Table S2: Critical quality attributes (CQAs) of AmB-loaded ODFs. Table S3: Co-efficient values and statistical parameters obtained for first order equations for the studied response variables: 1-Type of dextrose-derived-polymer film former, 2-Taste masking agent, 3-Type of Avicel, 4-Amount of Avicel, 5-Amount of plasticisers, 6-Amount of methanol, 7-Amount of cellulose-derived film formers. Results were analysed using a first order equation ($Y = B_0 + B_1X_1 + B_2X_2 + B_3X_3 + B_4X_4 + B_5X_5 + B_6X_6 + B_7X_7$) generated for the response variables investigated in the DoE. Seven coefficients (B1 to B7) were calculated with B0 as the intercept. Only those coefficients which were significant were retained in the simplified equations.

**Author Contributions:** Conceptualization, D.R.S. and A.L.; methodology: D.R.S. and A.L.; formal analysis, D.R.S., R.F.-G., M.M. and A.L.; resources, A.M.H. and A.L.; software: A.M.H. and A.L.; supervision: D.R.S. and A.L.; validation: A.L.; writing—original draft preparation, D.R.S., A.M.H. and A.L.; writing—review and editing, D.R.S., A.M.H. and A.L., supervision, D.R.S. and A.L.; project administration, A.L.; funding acquisition, A.M.H. and A.L.

**Funding:** This work was supported by a Sir Haley Stewart Trust grant (127) and a University of Portsmouth, Research and Development Fund to A. Lalatsa and a Science Foundation Ireland grant co-funded under the European Regional Development Fund (SFI/12/RC/2275) provided to A. M. Healy.

**Conflicts of Interest:** The authors declare no conflict of interest.

## References


1. Ruiz, H.K.; Serrano, D.R.; Dea-Ayuela, M.A.; Bilbao-Ramos, P.E.; Bolas-Fernandez, F.; Torrado, J.J.; Molero, G. New amphotericin B-gamma cyclodextrin formulation for topical use with synergistic activity against diverse fungal species and Leishmania spp. *Int. J. Pharm.* **2014**, *473*, 148–157. [CrossRef] [PubMed]
2. Montaga, M.T.; Lovero, G.; Coretti, C.; de Giglio, O.; Mratinelli, D.; Bedini, A.; Delia, M.; Rosato, A.; Codeluppi, M.; Caggiano, C. In vitro activities of amphotericin B deoxycholate and liposomal amphotericin B against 604 clinical yeast isolates. *J. Med. Microbiol.* **2014**, *63*, 1638–1643. [CrossRef] [PubMed]
3. Lagerlof, F.; Dawes, C. The volume of saliva in the mouth before and after swallowing. *J. Dent. Res.* **1984**, *63*, 618–621. [CrossRef] [PubMed]
4. Technical Data Sheet of Fungizone. Available online: https://botplusweb.portalfarma.com/documentos/2003/9/22/F46773%20FUNGIZONA%20IV%20200111.pdf (accessed on 15 January 2019).
5. AmBisome Technical Data Sheet. Available online: https://www.gilead.com/-/media/files/pdfs/medicines/other/ambisome/ambisome_pi.pdf (accessed on 14 March 2019).
6. Reiner, V.; Giarratana, N.; Monti, N.C.; Breitenbach, A.; Klaffenbach, P. Rapidfilm: An innovative pharmaceutical form designed to improve patient compliance. *Int. J. Pharm.* **2010**, *393*, 55–60. [CrossRef] [PubMed]
7. Gijare, C.; Deshpande, A. Orodispersible Films: A Systematic Patent Review. *Recent Patents Drug Deliv. Formul.* **2018**, *12*, 110–120. [CrossRef] [PubMed]
8. Irfan, M.; Rabel, S.; Bukhtar, Q.; Qadir, M.I.; Jabeen, F.; Khan, A. Orally disintegrating films: A modern expansion in drug delivery system. *Saudi Pharm. J.* **2016**, *24*, 537–546. [CrossRef]
9. Choudhary, D.R.; Patel, V.A.; Chhalotiya, U.K.; Patel, H.V.; Kundawala, A.J. Development and characterization of pharmacokinetic parameters of fast-dissolving films containing levocetirizine. *Sci. Pharm.* **2012**, *80*, 779–787. [CrossRef]
10. Dhiman, M.K.; Dhiman, A.; Sawant, K.K. Transbuccal delivery of 5-fluorouracil: Permeation enhancement and pharmacokinetic study. *AAPS Pharm. Sci. Technol.* **2009**, *10*, 258–265. [CrossRef]
11. Rolon, M.; Serrano, D.R.; Lalatsa, A.; de Pablo, E.; Torrado, J.J.; Ballesteros, M.P.; Healy, A.M.; Vega, C.; Coronel, C.; Bolas-Fernandez, F.; et al. Engineering Oral and Parenteral Amorphous Amphotericin B Formulations against Experimental Trypanosoma cruzi Infections. *Mol. Pharm.* **2017**, *14*, 1095–1106. [CrossRef]
12. Okamoto, H.; Taguchi, H.; Iida, K.; Danjo, K. Development of polymer film dosage forms of lidocaine for buccal administration. I. Penetration rate and release rate. *J. Control. Release* **2001**, *77*, 253–260. [CrossRef]
13. Senthilkumar, K.; Vijaya, C. Formulation Development of Mouth Dissolving Film of Etoricoxib for Pain Management. *Adv. Pharm.* **2015**, *2015*, 1–11. [CrossRef]
14. Serrano, D.R.; Walsh, D.; O'Connell, P.; Mugheirbi, N.A.; Worku, Z.A.; Bolas-Fernandez, F.; Galiana, C.; Dea-Ayuela, M.A.; Healy, A.M. Optimising the in vitro and in vivo performance of oral cocrystal formulations via spray coating. *Eur. J. Pharm. Biopharm.* **2018**, *124*, 13–27. [CrossRef] [PubMed]
15. Serrano, D.R.; Persoons, T.; D'Arcy, D.M.; Galiana, C.; Dea-Ayuela, M.A.; Healy, A.M. Modelling and shadowgraph imaging of cocrystal dissolution and assessment of in vitro antimicrobial activity for sulfadimidine/4-aminosalicylic acid cocrystals. *Eur. J. Pharm. Sci.* **2016**, *89*, 125–136. [CrossRef] [PubMed]
16. Serrano, D.R.; Hernandez, L.; Fleire, L.; Gonzalez-Alvarez, I.; Montoya, A.; Ballesteros, M.P.; Dea-Ayuela, M.A.; Miro, G.; Bolas-Fernandez, F.; Torrado, J.J. Hemolytic and pharmacokinetic studies of liposomal and particulate amphotericin B formulations. *Int. J. Pharm.* **2013**, *447*, 38–46. [CrossRef] [PubMed]
17. Torrado, J.J.; Espada, R.; Ballesteros, M.P.; Torrado-Santiago, S. Amphotericin B formulations and drug targeting. *J. Pharm. Sci.* **2008**, *97*, 2405–2425. [CrossRef] [PubMed]
18. Clinical and Laboratory Standards Institute. Method for antifungal disk diffusion susceptibility testing of yeast. In *CLSI document M44-A2*, 2nd ed.; Clinical and Laboratory Standards Institute: Wayne, PA, USA, 2009; Available online: https://clsi.org/media/1634/m44a2_sample.pdf (accessed on 7 July 2019).
19. Markande, A.; Nezzal, A.; Fitzpatrick, J.J.; Aerts, L. Investigation of the Crystallization Kinetics of Dextrose Monohydrate Using In Situ Particle Size and Supersaturation Monitoring. *Part. Sci. Technol.* **2009**, *27*, 373–388. [CrossRef]
20. Serrano, D.R.; Ballesteros, M.P.; Schatzlein, A.G.; Torrado, J.J.; Uchegbu, I.F. Amphotericin B Formulations —The Possibility of Generic Competition. *Pharm. Nanotechnol.* **2013**, *1*, 250–258. [CrossRef]

21. Rajagopalan, N.; Dicken, C.M.; Ravin, L.J.; Sternson, L.A. A study of the solubility of amphotericin B in nonaqueous solvent systems. *J. Parenter. Sci. Technol. Publ. Parenter. Drug Assoc.* **1988**, *42*, 97–102.
22. Pycia, K.; Gryszkin, A.; Berski, W.; Juszczak, L. The Influence of Chemically Modified Potato Maltodextrins on Stability and Rheological Properties of Model Oil-in-Water Emulsions. *Polymers* **2018**, *10*, 67. [CrossRef] [PubMed]
23. Iorgulescu, G. Saliva between normal and pathological. Important factors in determining systemic and oral health. *J. Med. Life* **2009**, *2*, 303–307. [PubMed]
24. Jun, S.; Zhang, H.; Bechhoefer, J. Nucleation and growth in one dimension. I. The generalized Kolmogorov-Johnson-Mehl-Avrami model. *Phys. Rev. E* **2005**, *71*, 011908. [CrossRef] [PubMed]
25. Poletto, M.; Pistor, V.; Zeni, M.; Zattera, A.J. Crystalline properties and decomposition kinetics of cellulose fibers in wood pulp obtained by two pulping processes. *Polym. Degrad. Stab.* **2011**, *96*, 679–685. [CrossRef]
26. Pistor, V.; Soares, B.G.; Mauler, R.S. Influence of the Polyhedral Oligomeric Silsesquioxane n-Phenylaminopropyl—POSS in the Thermal Stability and the Glass Transition Temperature of Epoxy Resin. *Polimeros* **2013**, *23*, 328–331. [CrossRef]
27. Borges, A.F.; Silva, C.; Coelho, J.F.; Simoes, S. Oral films: Current status and future perspectives II Intellectual property, technologies and market needs. *J. Control. Release* **2015**, *206*, 108–121. [CrossRef] [PubMed]
28. Lyu, X.; Zhao, C.; Yan, Z.M.; Hua, H. Efficacy of nystatin for the treatment of oral candidiasis: A systematic review and meta-analysis. *Drug Des. Dev. Ther.* **2016**, *10*, 1161–1171. [CrossRef] [PubMed]
29. Johnson, E.M.; Ojwang, J.O.; Szekely, A.; Wallace, T.L.; Warnock, D.W. Comparison of in vitro antifungal activities of free and liposome-encapsulated nystatin with those of four amphotericin B formulations. *Antimicrob. Agents Chemother.* **1998**, *42*, 1412–1416. [CrossRef] [PubMed]
30. Espada, R.; Valdespina, S.; Alfonso, C.; Rivas, G.; Ballesteros, M.P.; Torrado, J.J. Effect of aggregation state on the toxicity of different amphotericin B preparations. *Int. J. Pharm.* **2008**, *361*, 64–69. [CrossRef] [PubMed]
31. Espada, R.; Valdespina, S.; Molero, G.; Dea, M.A.; Ballesteros, M.P.; Torrado, J.J. Efficacy of alternative dosing regimens of poly-aggregated amphotericin B. *Int. J. Antimicrob. Agents* **2008**, *32*, 55–61. [CrossRef]
32. Singh, S.; Jain, S.; Muthu, M.S.; Tiwari, S.; Tilak, R. Preparation and evaluation of buccal bioadhesive films containing clotrimazole. *AAPS Pharm. Sci. Technol.* **2008**, *9*, 660–667. [CrossRef]

*pharmaceutics*

MDPI

*Article*

# Increased Efficacy of Oral Fixed-Dose Combination of Amphotericin B and AHCC® Natural Adjuvant against Aspergillosis

**Alba Pérez-Cantero [1], Dolores R. Serrano [2], Patricia Navarro-Rodríguez [1], Andreas G. Schätzlein [4], Ijeoma F. Uchegbu [3], Juan J. Torrado [2] and Javier Capilla [1,*]**

1 Unitat de Microbiologia, Facultat de Medicina i Ciències de la Salut, Universitat Rovira i Virgili and Institut d'Investigació Sanitària Pere Virgili (IISPV), 43201 Reus, Tarragona, Spain
2 Department of Pharmaceutics and Food Technology and Instituto Universitario de Farmacia Industrial (IUFI), School of Pharmacy, University Complutense, Avenida Complutense, 28040 Madrid, Spain
3 UCL School of Pharmacy, 29–39, Brunswick Square, London WC1N 1AX, UK
4 Nanomerics Ltd., St. Albans AL1 1SR, UK
* Correspondence: javier.capilla@urv.cat; Tel.: +34-977-759359; Fax: +34-977-759322

Received: 3 June 2019; Accepted: 19 August 2019; Published: 3 September 2019

**Abstract:** Invasive pulmonary aspergillosis represents one of the most serious fungal infections among immunocompromised patients. In this study, we aimed to analyze the in vivo efficacy of prophylactic oral amphotericin B (AMB) encapsulated in modified chitosan-nanoparticles (Nanomerics' Molecular Envelope Technology (MET)) supplemented with a standardized extract of cultured *Lentinula edodes* mycelia (AHCC®) in a murine model of pulmonary aspergillosis. We determined fungal burden and survival of mice and additionally, we carried out a cytokine analysis in an attempt to understand the immunomodulation of the extract. Our results evidenced equivalent efficacy between orally administered AMB-MET and the intravenous liposomal AMB marketed formulation. Addition of the AHCC® supplement significantly improved efficacy in terms of burden reduction and survival increase of both oral and intravenous AMB therapies compared to the untreated control group. Moreover, a protective effect of the extract was observed in terms of weight loss. Regarding the cytokine profiles, the Th1 immune response was stimulated in treated animals when compared to the control group. This response was marked by an enhancement in the MCP-1, GM-CSF, VEGF, RANTES and IL-17 levels and a decrease in the IL-6, a biomarker related to the severity of the infection.

**Keywords:** aspergillosis; amphotericin B; oral delivery; chitosan; shiitake; *Lentinula edodes*; AHCC®; Molecular Envelope Technology

## 1. Introduction

Invasive aspergillosis (IA) is a life-threatening condition that mostly affects immunocompromised patients, including those following stem cell or solid organ transplantation, patients with hematological malignancies, granulomatous disease, and neutropenia, or those under chemotherapy or corticosteroid treatment [1,2]. Within the genus, *Aspergillus fumigatus* is the main causal agent of the condition, with an associated mortality rate that ranges from 50 to 100% depending on the delay in diagnosis and treatment, as well as the underlying state of the patient or the affected organs [3,4].

The current IA treatment guidelines [5] establish voriconazole (VRC) as the treatment of choice, although the increase in the number of azole-resistant isolates observed in recent years has had a direct effect on patient outcomes, leading to therapeutic failure. Other recommended therapies include intravenous liposomal amphotericin B (LAMB), echinocandins or isavuconazole [5–8], yet other options are being studied in order to achieve better therapeutic outcomes and provide expanded alternatives of treatment. This has resulted in antifungal combinations [9,10], the development of new antifungal agents or improved formulations of the preexisting ones [11].

Amphotericin B (AMB) represented the first-line treatment for IA for many years due to its higher efficacy in treating aspergillosis. However, it is well-known that it can display great nephrotoxicity and cause a broad variety of side effects [12]. These complications are responsible for treatment interruption, which can exacerbate mortality rates or extend patient hospitalization [13–15]. Different AMB formulations are currently available for their parenteral use, including LAMB, AMB lipid complex (ABLC) or AMB colloidal dispersion (ABCD), all of them with reduced toxic effects when compared to the conventional micellar deoxycolate AMB formulation (dAMB). Even though these lipidic formulations have improved the efficacy-safety profile of the drug, they still require parenteral administration and, in addition, there are no oral formulations commercially available nowadays, which hampers amphotericin's use in developing countries [16].

In this context, novel AMB formulations are currently under research in order to achieve safer formulations that could be orally administered and, therefore, improve the patients' quality of life. Some of these in-development formulations include nanoparticle-based carrier systems [17–19]. In this sense, we demonstrated in a previous work the efficacy of oral AMB encapsulated in *N*-palmitoyl-*N*-monomethyl-*N*,*N*-dimethyl-*N*,*N*,*N*-trimethyl-6-*O*-glycol chitosan nanoparticles (AMB-MET) in a murine model of invasive aspergillosis, in which we achieved translocation to specific organs of the pathology, sparing the site of toxicity with a comparable efficacy to LAMB [15].

Despite the advances in the development of new drug delivery systems, the resolution of IA remains challenging, in part due to the immunocompromised state of the patients. Therefore, boosting the immune status of the patients seems a key point in order to satisfactorily treat IA. The use of natural extracts as potential adjuvants that could stimulate and enhance the immune system is emerging as a promising approach for the treatment of different illnesses. For instance, *Lentinula edodes*, the shiitake mushroom, has been traditionally used to treat quotidian aches and aging fatigue [20]. Other properties associated with this mushroom are related to heart health or lung diseases, however, there is an important lack of experimental data [20]. Nevertheless, shiitake extract's benefits have been demonstrated in mice, as they have been used as adjuvants in hepatitis vaccines [21]. Additionally, chemotherapy-subjected patients have displayed longer survival when shiitake extract has been combined with conventional chemotherapy [22].

One of the most studied extracts of the shiitake mushroom is AHCC® [23]. AHCC® is a standardized cultured extract of shiitake mycelia, manufactured by a proprietary mycelia cell culture process, as its components differ according to the culture conditions [24]. In particular, AHCC® is mainly composed of carbohydrates such as acetylated α-1,4 glucan, but also contains proteins, minerals, fats, and fiber [25], and it has been suggested to play an important role in the orchestration of the immune response to viral, parasitic, and bacterial infections, as well as the maintenance of immune system homeostasis [26]. Consequently, the aim of the present study was to evaluate the in vivo efficacy of oral AMB-MET supplemented with AHCC® in a murine model of invasive pulmonary aspergillosis as a potential prophylactic therapy.

## 2. Materials and Methods

### 2.1. Materials

The AMB was purchased from Azelis (Barcelona, Spain). The AHCC® was kindly provided by Amino Up Co., Ltd. (Sapporo, Japan). The AmBisome® was supplied by Gilead Sciences S.L. (Madrid, Spain). The sodium deoxycholate was supplied by Fluka Chemie AG (Buchs, Switzerland). All other reagents and chemicals were obtained from Sigma Aldrich Chemical Co. (Poole, UK or Madrid, Spain), Panreac S.A. (Barcelona, Spain), Fisher Scientific (Loughborough, UK) or VWR (Lutterworth, UK). The AMB-MET nanoparticles were prepared and characterized as previously reported [15].

### 2.2. Animals

A total of 75 OF-1 four-week-old male mice weighing 35 g (Charles River, Criffa S.A. Barcelona, Spain) were divided into groups of 15 animals/group, with 10 used for the survival studies and 5 for fungal load determination. Animals were housed under standard conditions with water and food ad libitum. All procedures were supervised and approved by the Universitat Rovira i Virgili Animal Welfare and Ethics Committee (URV reference 0247, 15 March 2017).

### 2.3. Inoculum

One *A. fumigatus* clinical strain (FMR 7738) isolated from blood was cultured at 37 °C for 5 days on potato dextrose agar (PDA) plates prior to its use. The inoculum was prepared by flooding the culture plate with 5 mL of saline with 0.05% Tween® 20, with the application of gentle agitation to release the conidia. Afterward, the conidial suspension was serially diluted and adjusted to the desired concentration by a hemocytometer count.

### 2.4. Murine Model of Pulmonary Aspergillosis

A murine model of pulmonary aspergillosis was used as previously described [27]. Immunosuppression of mice was established 4 days before infection and every 3 days by subcutaneous injection of cortisone acetate at 125 mg/kg [28]. On the day of infection, mice were anesthetized with inhaled sevoflurane and inoculated by nasal instillation with 25 μL of a suspension containing $5 \times 10^4$ conidia. Animals were checked twice daily until the end of the experiment (11 days post-infection).

### 2.5. Treatment

Animals received prophylactic treatment starting 5 days before infection, followed by therapeutic treatment. Prophylaxis consisted of AMB-MET administered orally by gavage (p.o) at 2.5 mg/kg twice a day (BID) or intravenous (i.v.) LAMB at 2.5 mg/kg BID. Both drugs were administered alone or in combination with AHCC® administered p.o. twice a day at 270 mg/kg. On the day of infection, no compound was administered. One day after infection and for 11 days, the animals received oral AMB-MET at 5 mg/kg p.o. BID or LAMB at 5 mg/kg i.v. BID, both alone or in combination with oral AHCC® administered BID at 270 mg/kg. The control animals received only the vehicle (PBS).

### 2.6. Fungal Burden Studies

Six days after the infection, the animals from the tissue burden group were anesthetized by inhaled sevoflurane. The blood was obtained by cardiac puncture and the serum was collected after blood centrifugation and frozen until used for further analysis. In addition, 11 days post-infection, the surviving mice from the mortality study groups were also subjected to cardiac puncture in the same manner. In all cases, immediately after blood collection, the animals were euthanized by cervical dislocation, and brain, kidneys, and lungs were aseptically removed, weighed, and mechanically homogenized in 1 mL of sterile saline. Homogenates were serially 10-fold diluted, placed onto PDA plates and incubated at 35 °C for Colony Forming Unit (CFU) determination.

### 2.7. Sera Determinations

The AMB levels were determined from all obtained sera by bioassay as previously described [29] using *Candida parapsilosis* ATCC 22019 as the control strain, but also by a validated HPLC method [30]. Briefly, the AMB was isocratically eluted using a Thermo® Hypersil BDS C18 reverse-phase column (200 × 4.6 mm, 5 μm) and a mobile phase consisted of acetonitrile, acetic acid, and water (52:4.3:43.7, *v*/*v*/*v*) with a flow rate of 1 mL $min^{-1}$. The absorbance was monitored at 406 nm and the injection volume was set at 40 μL. The cytokine quantification was performed from the animal sera samples consisting of 100 μL from the control (6 and 11 days post-infection) and the treatment groups (11 days post-infection) using a Quantibody® array (RayBiotech, Norcross, GA, USA, distributed by BioNova cientifica, Madrid, Spain). This multiplexed sandwich ELISA-based quantitative array platform was able to analyze: GM-CSF (granulocyte-macrophage colony-stimulating factor), IFN-$\gamma$ (interferon), IL-1a, IL-1b, IL-2, IL-3, IL-4, IL-5, IL-6, IL-9, IL-10, IL-12, IL-13, IL-17, KC (chemokines), MCP (monocyte chemoattractant protein)-1, M-CSF (macrophage colony-stimulating factor), TNF-$\alpha$ (tumor necrosis factor), RANTES (regulated on activation, normal T cell expressed and secreted), and VEGF (vascular endothelial growth factor) using a pair of cytokine specific antibodies for detection. A capture antibody was first bound to the glass surface and after incubation with the sample, the target cytokine was trapped on the solid surface. A second biotin-labeled detection antibody was then added, which could recognize a different epitope of the target cytokine. The cytokine-antibody-biotin complex was visualized through the addition of the streptavidin-conjugated Cy3 equivalent dye using the GenePix® 4000B Scanner (Unidad de Genomica, Campus Moncloa, UCM, Madrid, Spain).

### *2.8. Statistical Analysis*

The fungal burden results were analyzed by the Mann–Whitney U test and the survival curves were compared among them by the log-rank test using GraphPad Prism 6.0 for Windows (GraphPad Software Inc; La Jolla, CA, USA).

The statistical differences in the sera AMB levels were evaluated via a one-way ANOVA test using Minitab 15 (Minitab Ltd., Coventry, UK), Tukey's test was used for paired-group comparisons.

The cytokine statistical analysis was performed using both a univariate and multivariate technique as previously reported [31]. The normality test (Shapiro–Wilk) was performed. The Kruskal–Wallis pairwise comparison test was carried out when the $P$-value was $< 0.05$ while a one-way ANOVA (Tukey's post hoc test) was used for those cytokines that followed a normal distribution ($P > 0.05$). The multivariate data analysis (MVA) was performed using the Unscrambler® X software (CAMO Software, Oslo, Norway). The treatment effect on the cytokine profile was analyzed by principal component analysis (PCA) to study systematic variability and the relationships between variables and scores. The correlation loadings of the principal components (PCs) represented the variance for each variable for a given PC, giving information about the variability source inside the dataset. $P$-values $\leq 0.05$ were considered as statistically significant.

## 3. Results

### *3.1. Survival and Body Weight Progression*

The pulmonary infection resulted in elevated mortality in animals from the control group. The untreated animals began to die six days after challenge and at the end of the experiment (day 11 post-infection), only 30% had survived to the infection (Figure 1A). Oral AMB-MET or i.v. LAMB alone did not increase, in a significant manner, the survival rates in comparison to the control group ($P = 0.171$ and 0.130, respectively). However, a statistically significant beneficial effect was observed in terms of survival when oral AMB-MET and i.v. LAMB were administered in combination with oral AHCC® ($P = 0.033$ and 0.029, respectively, in comparison to the control group). No significant differences were observed between the LAMB-AHCC® and AMB-MET-AHCC® treatments ($P = 0.530$).

Regarding the animals' weight, which was checked daily starting five days before the infection and until the end of the experiment, the results showed a considerable decrease (15%) in the untreated mice (Figure 1B). The LAMB-treated mice were the group with the most significant weight loss (19.2%), while the supplementation of the treatment with AHCC® only resulted in an 11% reduction. The AMB-MET and AMB-MET-AHCC® were the treatments associated with a minor weight reduction, with an 8.5% and 9.5% average weight loss ($P \leq 0.0068$), respectively, indicating good gastrointestinal tolerance after oral administration.

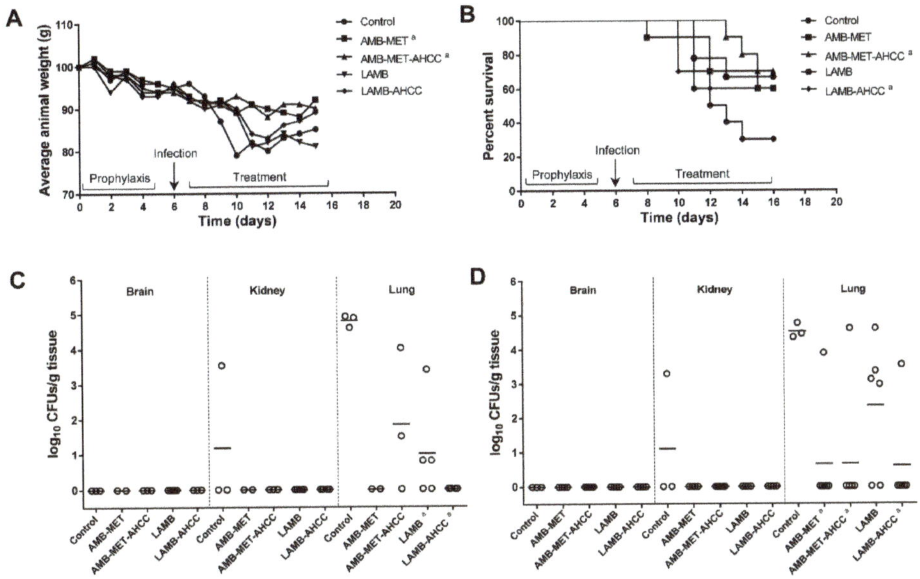


**Figure 1.** Effect of AMB-MET p.o. and LAMB i.v. with and without AHCC® supplementation in immunosuppressed mice infected intranasally with *Aspergillus fumigatus*. (**A**) Progression of body weight. (**B**) Percent of surviving mice. Panels (**C**,**D**) show scattergram and median (horizontal lines) of CFUs recovered from targeted organs at 6 and 11 days after infection, respectively. [a] Indicates statistically significant differences ($P < 0.05$) in comparison to the control group.

### 3.2. Fungal Burden

The tissue burden study performed six days post-infection showed poor dissemination of *Aspergillus* to the brain and kidneys in control animals, while a high amount of fungus was recovered from the lung tissue (4.80 ± 0.17 $\log_{10}$ CFUs/g) (Figure 1C). The AMB-MET and LAMB monotherapies reduced the CFUs from the lungs (0 and 1.00 ± 0.46 $\log_{10}$ CFUs/g, respectively), in a significant manner with respect to the control animals (both $P = 0.018$). Similarly, the addition of AHCC® reduced the fungal load from the lungs in those animals receiving LAMB ($P = 0.0286$), although this significant effect was not observed in those treated with AMB-MET ($P = 0.06$) in comparison to the control group. No significant differences in the tissue burden reduction were found between the monotherapies ($P = 0.4286$) or between the two AHCC®-combined therapies ($P = 0.1429$).

By day 11 post-infection, no fungal loads were detected in the brain of the control or treated groups of animals, and only one kidney from the control group was positive for fungal presence. The lungs remained highly affected in the untreated animals (4.52 ± 0.21 $\log_{10}$ CFUs/g) (Figure 1D). At this time, LAMB therapy was unable to significantly reduce the tissue burden when compared to the controls ($P = 0.083$). However, the combination of LAMB with AHCC® resulted in a great lung burden reduction ($P = 0.0119$) with 83% of animals displaying infection clearance. Similarly, high clearance was obtained in the animals receiving AMB-MET with or without AHCC® (86% and 83% infection clearance, respectively) and this reduction was statistically significant in comparison with the untreated group ($P = 0.0119$ and 0.0250, respectively). No statistically significant differences were observed among the various treatment groups ($P \geq 0.197$).

### 3.3. AMB Concentration in Plasma

The AMB levels in plasma were significantly higher for those animals treated with LAMB compared to AMB-MET ($P < 0.05$) (Figure 2). The administration of AHCC® as an immunoadjuvant did not alter in a significant manner the AMB concentration in the plasma in any of the formulations. Additionally, no differences were observed in the AMB plasma levels in those animals treated with AMB-MET at day six and day 11 post-infection. However, the AMB levels were significantly greater at day 11 compared to day six for LAMB.

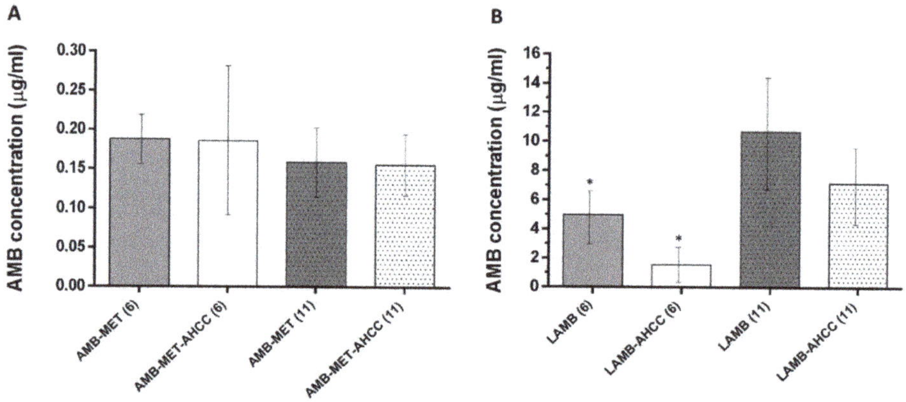


**Figure 2.** AMB concentration in plasma after day 6 (6) and day 11 (11) post-infection. (**A**) Animals treated with AMB-MET with and without AHCC®. (**B**) Animals treated with LAMB with and without AHCC®. * $P$-value< 0.05 Day 6 vs. day 11 post-infection.

### *3.4. Cytokine Analysis*

In general, all treatments showed the tendency to increase the production of the cytokines IFN-$\gamma$, TNF-$\alpha$, GM-CSF, RANTES, VEFG, IL-4, IL-17, IL-2, and especially MCP-1 ($P$ = 0.002) and IL-1b ($P$ = 0.043), while the rest of the cytokines were diminished or maintained in comparison to the control group (Figures 3 and 4). In the case of the animals treated with LAMB, greater levels of GM-CSF, MCP-1 and VEGF were found compared to the control group, while administration of AMB-MET led to higher secretion of IL-1b, IL-12, and IL-17. The combination of LAMB-AHCC® increased the levels of IL-2 and IL-12, while in those receiving AMB-MET the administration of AHCC® increased RANTES and VEFG. AHCC® administered in combination with LAMB or AMB-MET diminished the release of IL-1b, IL-4, IL-5, IL-6, and IL-9 in comparison to the monotherapies. It is worth mentioning the slight reduction of IL-6 in the treated animals and especially in those receiving AHCC®, since IL-6 is a biomarker of the infection severity ($P$ = 0.059). The secretion of IL-12, IL-17, TNF-$\alpha$, and GM-CSF was higher in those animals receiving AMB-MET-AHCC® than in the animals treated only with AMB-MET. Contrarily, these cytokines were detected at equal or slightly higher concentrations in animals treated with LAMB-AHCC® when compared to the LAMB group.

The PCA (Figure 5) showed that most of the variance found in the cytokine analysis was due to two principal components (84% for PC1 and 7% for PC2). When the samples were grouped based on the PCA, the majority of controls were grouped together (blue color) while most of the treated animals were grouped within each other (red color). However, some differences were observed, as the other two groups (green and brown) were also differentiated and shared common features (Figure 5A). Regarding the correlation loading plots, the IL-6 response was located completely opposed to the markers of a positive prognosis, such as RANTES, IL-2 or IFN-$\gamma$ (Figure 5B). Additionally, an antagonistic effect was observed between the responses of VEGF and MCP-1 (higher with the LAMB treatment) versus Il-12 and IL-17 (enhanced in the AMB-MET group).

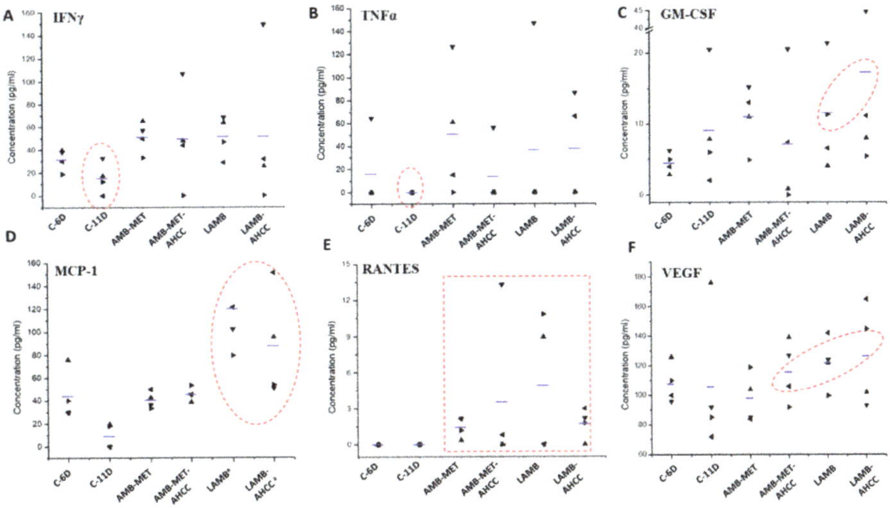


**Figure 3.** Scattergram of cytokine levels at day 6 and 11 for control group and at day 11 post-infection for all treatment groups. Horizontal lines indicate mean values of (**A**) IFNγ, (**B**)TNFα, (**C**) GM-CSF, (**D**) MCP-1, (**E**) RANTES, (**F**) VEGF. Key: C-6D, control group at day six post-infection, C-11D, control group at day 11 post-infection, AMB-MET, and AMB-MET-AHCC, treatment with AMB encapsulated within GCPQ nanoparticles alone or in combination with AHCC® respectively, LAMB and LAMB-AHCC, treatment with i.v. liposomal AMB alone or in combination with AHCC® respectively. The most relevant changes in the cytokine patterns have been highlighted within a red circle. [a] $P < 0.05$.

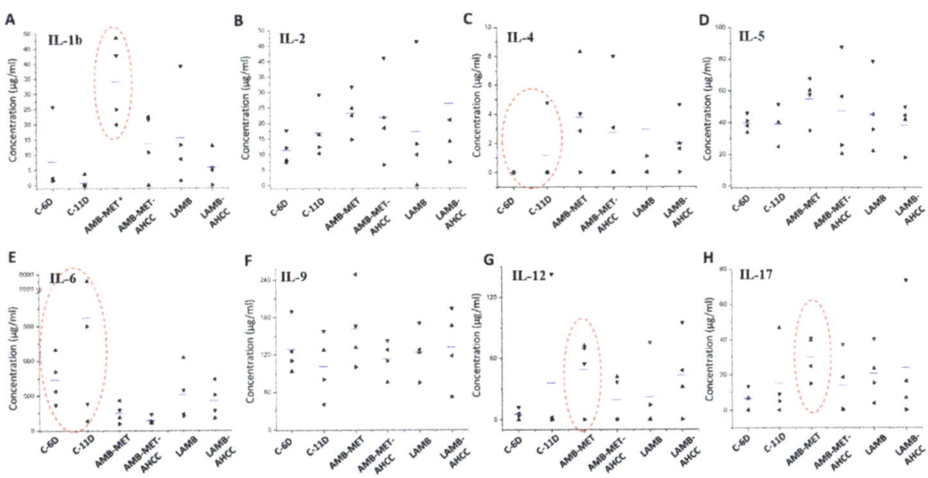


**Figure 4.** Scattergram of cytokine at day 6 and 11 for control group and at day 11 post-infection for all treatment groups. Horizontal lines indicate mean values of (**A**) IL-1b, (**B**) IL-2, (**C**) IL-4, (**D**) IL-5, (**E**) IL-6, (**F**) IL-9, (**G**) IL-12, (**H**) IL-17. Key: C-6D, control group at day six post-infection, C-11D, control group at day 11 post-infection, AMB-MET, and AMB-MET-AHCC, treatment with AMB encapsulated within GCPQ nanoparticles alone or in combination with AHCC® respectively, LAMB and LAMB-AHCC, treatment with iv liposomal AMB alone or in combination with AHCC® respectively. The most relevant changes in the cytokine patterns have been highlighted within a red circle. [a] $P < 0.05$.

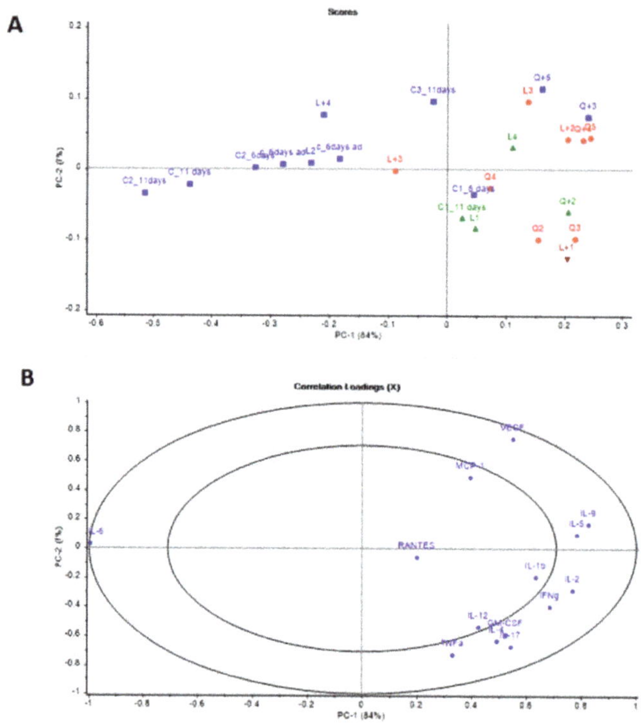


**Figure 5.** Principal component analysis of the cytokine profile at day six and 11 for control group and at day 11 post-infection for all treatment groups. (**A**) Score plot. (**B**) Correlation loading plot. Key: C represents animals from the untreated control group; L- animals treated with LAMB and Q- animals treated with AMB-MET.

## 4. Discussion

The discovery of new options for the treatment of IA is essential in order to overcome the therapeutic failure associated with this condition. Therapeutic failure is mostly due to the increase of azole resistant *Aspergillus* isolates and the toxicity events associated with the current therapies [32,33]. With the difficulties and limitations linked to the development of new antifungal drugs and the finding of new safe-to-human targets, establishing new therapeutic and prophylactic approaches with the already available drugs is crucial.

In this study, we aimed to test the in vivo efficacy of a recently engineered chitosan encapsulated AMB nanoformulation, AMB-MET [15], supplemented with the natural extract AHCC® from the mushroom *Lentinula edodes* as prophylactic treatment in order to clarify and characterize its role as an adjuvant. The supplementation with AHCC® has already shown benefits in the treatment of melanoma in animal models [34]. Other researchers have also found similar effects when administering AHCC® alone or in combination with drugs destined to treat other conditions, such as gastric cancer, but also when combined with their chemotherapy treatment in murine models [22]. In the field of infectious diseases, different studies have demonstrated a beneficial effect of AHCC®, specifically against canine leishmaniosis [35], influenza [36], *Chlamydia trachomatis* infections [37], and West Nile encephalitis in mouse models [38]. In addition, the positive effects against bacterial infections have been described [39].

Regarding our study, the in vivo efficacy of the treatments we assayed (AMB-MET and LAMB) against aspergillosis has already been established by others [15,40,41]. Nevertheless, the results we found in terms of survival in all of the treated animals revealed a significantly higher efficacy of the assayed drugs when supplemented with the natural extract AHCC®, in comparison to the untreated group, which evidences a clear beneficial effect of the natural extract. As we also compared the fungal burden of the treated and untreated animals at days six and 11 post-infection, we found a burden reduction in the lung tissue of all the assayed treatments at both experimentation points. Surprisingly, at day six, the AMB-MET+AHCC®, LAMB, and LAMB+AHCC® treatments reduced the burden, in a significant manner. At day 11, this was achieved by AMB-MET, AMB-MET+AHCC®, and LAMB+AHCC®, suggesting that AMB-MET may have a wider activity in the long-term. This could be due to the preferential distribution in the lung tissue after oral administration of AMB-MET [15] and the higher toxicity of i.v. LAMB when administered for prolonged periods of time, which might have also led to a significant weight loss in the treated animals of this group. Along with the AHCC® adjuvant role on efficacy, we found that this extract acted as a protective agent of weight loss and appetite diminution, which is very well exemplified in the weight loss differences between the LAMB and LAMB+AHCC® therapies, accompanied by an average weight loss of 19.2% and 10%, respectively. Several nutritional effects have already been linked to this kind of glucans, such as the reduction of stress and cholesterol levels, hypoglycemic effects or improvement of ulcerative colitis cases [42].

Besides, the protective role of AHCC® in infectious diseases seems to be related to its immunomodulatory properties by priming TLR-4 and TLR-2 [26] and the associated TLR-4/MyD88 and NF-κB/MAPK signal transduction pathways [43]. In this sense, an increase in MCP-1 secretion by intestinal epithelial cells and an increase in both IL-1b and TNFα secretion by monocytes have been reported as the key effect after administration of AHCC® [43]. In addition, AHCC® has previously been shown to potentially promote macrophage and natural killer cell proliferation [43,44]. On this basis, we performed a cytokine analysis to better characterize the immunomodulatory effects of AHCC® in our treatment groups against aspergillosis. A schematic representation of the immunological response associated to *A. fumigatus* infections is illustrated in Figure 6.

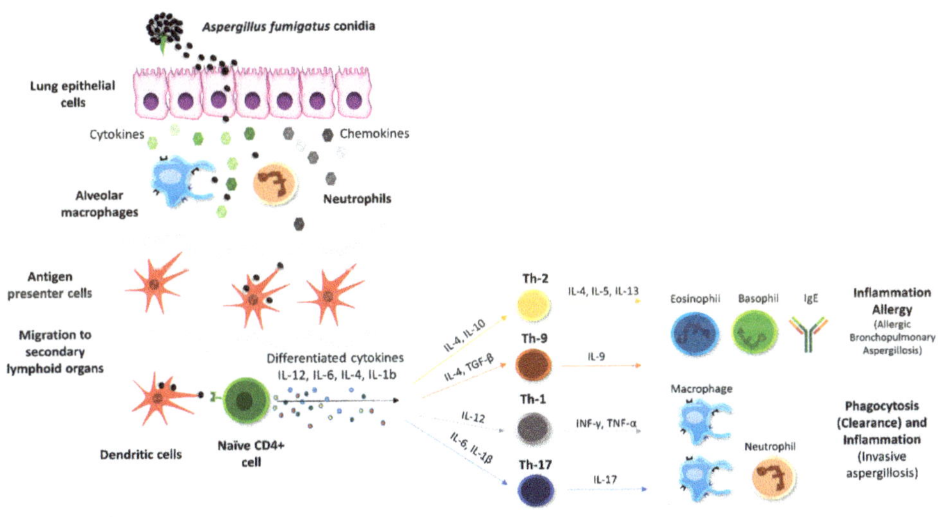


**Figure 6.** Schematic representation of immunological response against *A. fumigatus* antigens and production of differentiated cytokines and effector $T_H$-cells (adapted and modified from [45]).

In the case of fungal infections, the outcome partly depends on the Th1 protective cellular response, which is principally driven by the proinflammatory cytokines TNF, IFN, IL-6, IL-12, and IL-1. In contrast, the Th2 response is not protective but regulates the excessive inflammatory conduction of Th1. This response is characterized by the increase in the cytokines IL-4, IL-5, IL-13, and the promotion of non-opsonizing antibodies. Other cytokines also related to fungal infections are IL-8, IL-10, and IL-15, as well as the already mentioned IL-17 [46–48]. Interestingly, in our work, most of the studied cytokines were found to be higher in those animals receiving any treatment in comparison to the control group, with the exception of IL-6, a biomarker related to the severity of the infection, which was accordingly reduced in the treated animals. Our results also showed a slight increase in the Th2 response in the treated animals that, combined with the increase of the Th17 response promoted by IL-17, can display a positive effect due to its anti-inflammatory and tissue protection properties [46].

Furthermore, increased levels of MCP-1, GM-CSF, and VEGF, which have been associated with favorable outcomes in infectious diseases by promoting monocyte attraction, macrophage maturation and killing [49–51], were also found in our treated groups. In addition, both the AMB-MET and LAMB treated animals showed an increase in IFN-$\gamma$ levels compared to the control group, which is correlated with a better prognosis since IFN-$\gamma$ activates the immune system response through the stimulation of macrophages and neutrophils [52]. Similar results were obtained for TNF-$\alpha$, even though only a few animals from the treated groups displayed a response for this biomarker. The increase in TNF-$\alpha$ is also associated with a more positive response towards microorganisms, since it stimulates the protective Th1 response [53]. Likewise, we observed increased levels of monocyte chemotactic protein (MCP-1), especially in the LAMB group, which has a favorable role in pulmonary infections due to its association with the attraction of more monocytes to the site of infection and the promotion and maturation to macrophages, and hence, MCP-1 facilitates microorganism eradication [49].

Another biomarker that can be used in systemic fungal infections is RANTES. RANTES levels, which drop drastically in sick individuals with systemic fungal infections [50], were reestablished in our treated animals, while its levels were correspondingly undetectable in the control group.

In the case of the GM-CSF, its levels were slightly increased in the AMB-MET and LAMB groups, which was a positive response, considering that higher values of GM-CSF are associated with macrophage maturation, stimulation and, therefore, with improved innate immune response [49]. Finally, with regard to VEGF, slightly greater levels were found in the LAMB group compared to the other treatments. Promisingly, previous reports on pulmonary aspergillosis models have associated higher values of VEGF with greater animal survival [51]. We are aware of the limitations of the present study due to the low number of surviving animals in the sampling days (i.e., day six and 11 post-infection), which together with the high dispersion of data prevent us from robustly linking the measured cytokines levels with the observed benefit of using AHCC©.

Taken together, our data suggest that oral administration of AMB-MET matches the efficacy of i.v. liposomal AMB against *Aspergillus*, being of great advantage to be used for immunocompromised patients discharged from hospital. Moreover, the combination of AMB with AHCC®, with positive effects at six and 11 days post-infection, seems to play an extremely important role as an adjuvant by boosting the immune system response and protecting against weight loss and appetite diminution.

## 5. Conclusions

Our in vivo study results evidenced high efficacy against aspergillosis of AMB-MET and LAMB therapies when combined with the AHCC® supplement since, in contrast to the AMB-based monotherapies, AMB-MET/LAMB with AHCC® supplementation displayed statistically significant improvements with respect to the control group in terms of survival and burden reduction. Although the AHCC® immune system boost is very promising, further studies on its immunomodulation and its potential use as an adjuvant with other antifungals against invasive fungal infections need to be carried out.

**Author Contributions:** Conceptualization, A.G.S. and J.J.T.; Data curation, A.G.S. and I.F.U.; Formal analysis, A.P.-C., D.R.S., P.N.-R., J.J.T. and J.C.; Investigation, A.P.-C., D.R.S., P.N.-R., A.G.S., J.J.T. and J.C.; Methodology, I.F.U. and J.C.; Resources, I.F.U. and J.J.T.; Writing—original draft, A.P.-C.; Writing—review & editing, D.R.S., I.F.U. and J.C.

**Funding:** This research was funded by the Complutense University and Madrid Community Administration (research group 910939) and the UCL School of Pharmacy, University of London (reference: 00012151). A.P.-C. is the recipient of a FI fellowship from Generalitat de Catalunya (Spain).

**Acknowledgments:** We thank Amino Up Co., Ltd. (Sapporo, Japan) for the gift of AHCC®. We also thank Nanomerics Ltd. (London, UK) for the gift of Nanomerics' MET.

**Conflicts of Interest:** The authors declare no conflict of interest. Nanomerics Ltd. kindly provided Nanomerics' MET, however they displayed no role or participation in any section of this research article.

## References

1. Chowdhary, A.; Sharma, C.; Meis, J.F. Azole-Resistant Aspergillosis: Epidemiology, Molecular Mechanisms, and Treatment. *J. Infect. Dis.* **2017**, *216*, S436–S444. [CrossRef] [PubMed]
2. Gonçalves, S.S. Global Aspects of Triazole Resistance in Aspergillus fumigatus with Focus on Latin American Countries. *J. Fungi* **2017**, *3*, 5. [CrossRef]
3. Brown, G.D.; Denning, D.W.; Gow, N.A.R.; Levitz, S.M.; Netea, M.G.; White, T.C. Hidden killers: Human fungal infections. *Sci. Transl. Med.* **2012**, *4*, 165rv13. [CrossRef] [PubMed]
4. Lamoth, F. Aspergillus fumigatus-Related Species in Clinical Practice. *Front. Microbiol.* **2016**, *7*, 683. [CrossRef]
5. Patterson, T.F.; Thompson, G.R.; Denning, D.W.; Fishman, J.A.; Hadley, S.; Herbrecht, R.; Kontoyiannis, D.P.; Marr, K.A.; Morrison, V.A.; Nguyen, M.H.; et al. Practice guidelines for the diagnosis and management of aspergillosis: 2016 update by the infectious diseases society of America. *Clin. Infect. Dis.* **2016**, *63*, e1–e60. [CrossRef] [PubMed]
6. Wiederhold, N.; Patterson, T. Emergence of Azole Resistance in Aspergillus. *Semin. Respir. Crit. Care Med.* **2015**, *36*, 673–680. [CrossRef]
7. Rivero-Menendez, O.; Alastruey-Izquierdo, A.; Mellado, E.; Cuenca-Estrella, M. Triazole Resistance in Aspergillus spp.: A Worldwide Problem? *J. Fungi* **2016**, *2*, 21. [CrossRef]
8. Aruanno, M.; Glampedakis, E.; Lamoth, F. Echinocandins for the treatment of invasive aspergillosis: From laboratory to bedside. *Antimicrob. Agents Chemother.* **2019**, *63*, e00399-19. [CrossRef]
9. Martin-Vicente, A.; Capilla, J.; Guarro, J. In Vivo Synergy of Amphotericin B plus Posaconazole in Murine Aspergillosis. *Antimicrob. Agents Chemother.* **2016**, *60*, 296–300. [CrossRef]
10. Bedin, D.L.; de Jesus, F.P.K.; Keller, J.T.; Weiblen, C.; de Azevedo, M.I.; Oliveira, V.; Morais, S.J.; Hartz, A.S. Evaluation of the efficacy of a posaconazole and anidulafungin combination in a murine model of pulmonary aspergillosis due to infection with Aspergillus fumigatus. *Diagn. Microbiol. Infect. Dis.* **2018**, *90*, 40–43. [CrossRef]
11. Scorzoni, L.; de Paula e Silva, A.C.A.; Marcos, C.M.; Assato, P.A.; de Melo, W.C.M.A.; de Oliveira, H.C.; Costa-Orlandi, C.B.; Mendes-Giannini, M.J.; Fusco-Almeida, A.M. Antifungal Therapy: New Advances in the Understanding and Treatment of Mycosis. *Front. Microbiol.* **2017**, *8*, 36. [CrossRef] [PubMed]
12. Steimbach, L.M.; Tonin, F.S.; Virtuoso, S.; Borba, H.H.L.; Sanches, A.C.C.; Wiens, A.; Fernandez-Llimós, F.; Pontarolo, R. Efficacy and safety of amphotericin B lipid-based formulations—A systematic review and meta-analysis. *Mycoses* **2017**, *60*, 146–154. [CrossRef] [PubMed]
13. Falagas, M.E.; Karageorgopoulos, D.E.; Tansarli, G.S. Continuous versus Conventional Infusion of Amphotericin B Deoxycholate: A Meta-Analysis. *PLoS ONE* **2013**, *8*, e77075. [CrossRef] [PubMed]
14. Falci, D.R.; da Rosa, F.B.; Pasqualotto, A.C. Comparison of nephrotoxicity associated to different lipid formulations of amphotericin B: A real-life study. *Mycoses* **2015**, *58*, 104–112. [CrossRef] [PubMed]
15. Serrano, D.R.; Lalatsa, A.; Dea-Ayuela, M.A.; Bilbao-Ramos, P.E.; Garrett, N.L.; Moger, J.; Guarro, J.; Capilla, J.; Ballesteros, M.P.; Schätzlein, A.G.; et al. Oral particle uptake and organ targeting drives the activity of amphotericin B nanoparticles. *Mol. Pharm.* **2015**, *12*, 420–431. [CrossRef] [PubMed]
16. de Castro Spadari, C.; Lopes, L.B.; Ishida, K. Potential Use of Alginate-Based Carriers As Antifungal Delivery System. *Front. Microbiol.* **2017**, *8*, 97. [CrossRef]
17. Singh, P.K.; Jaiswal, A.K.; Pawar, V.K.; Raval, K.; Kumar, A.; Bora, H.K.; Dube, A.; Chourasia, M.K. Fabrication of 3-O-sn-Phosphatidyl-L-serine Anchored PLGA Nanoparticle Bearing Amphotericin B for Macrophage Targeting. *Pharm. Res.* **2018**, *35*, 60. [CrossRef]

18. Chaudhari, M.B.; Desai, P.P.; Patel, P.A.; Patravale, V.B. Solid lipid nanoparticles of amphotericin B (AmbiOnp): In vitro and in vivo assessment towards safe and effective oral treatment module. *Drug Deliv. Transl. Res.* **2015**, *6*, 354–364. [CrossRef]
19. Jain, S.; Valvi, P.U.; Swarnakar, N.K.; Thanki, K. Gelatin Coated Hybrid Lipid Nanoparticles for Oral Delivery of Amphotericin B. *Mol. Pharm.* **2012**, *9*, 2542–2553. [CrossRef]
20. Money, N.P. Are mushrooms medicinal? *Fungal Biol.* **2016**, *120*, 449–453. [CrossRef]
21. Zhang, Y.; Li, S.; Wang, X.; Zhang, L.; Cheung, P.C.K. Advances in lentinan: Isolation, structure, chain conformation and bioactivities. *Food Hydrocoll.* **2011**, *25*, 196–206. [CrossRef]
22. Ina, K.; Kataoka, T.; Ando, T. The use of lentinan for treating gastric cancer. *Anticancer Agents Med. Chem.* **2013**, *13*, 681–688. [CrossRef] [PubMed]
23. Xu, X.; Yan, H.; Tang, J.; Chen, J.; Zhang, X. Polysaccharides in *Lentinus edodes*: Isolation, Structure, Immunomodulating Activity and Future Prospective. *Crit. Rev. Food Sci. Nutr.* **2014**, *54*, 474–487. [CrossRef] [PubMed]
24. Sari, M.; Prange, A.; Lelley, J.I.; Hambitzer, R. Screening of beta-glucan contents in commercially cultivated and wild growing mushrooms. *Food Chem.* **2017**, *216*, 45–51. [CrossRef] [PubMed]
25. Hirose, A.; Sato, E.; Fujii, H.; Sun, B.; Nishioka, H.; Aruoma, O.I. The influence of active hexose correlated compound (AHCC) on cisplatin-evoked chemotherapeutic and side effects in tumor-bearing mice. *Toxicol. Appl. Pharmacol.* **2007**, *222*, 152–158. [CrossRef] [PubMed]
26. Mallet, J.F.; Graham, É.; Ritz, B.W.; Homma, K.; Matar, C. Active Hexose Correlated Compound (AHCC) promotes an intestinal immune response in BALB/c mice and in primary intestinal epithelial cell culture involving toll-like receptors TLR-2 and TLR-4. *Eur. J. Nutr.* **2016**, *55*, 139–146. [CrossRef] [PubMed]
27. Lepak, A.J.; Marchillo, K.; Vanhecker, J.; Andes, D.R. Posaconazole pharmacodynamic target determination against wild-type and Cyp51 mutant isolates of Aspergillus fumigatus in an in vivo model of invasive pulmonary aspergillosis. *Antimicrob. Agents Chemother.* **2013**, *57*, 579–585. [CrossRef] [PubMed]
28. Dixon, D.M.; Polak, A.; Walsh, T.J. Fungus dose-dependent primary pulmonary aspergillosis in immunosuppressed mice. *Infect. Immun.* **1989**, *57*, 1452–1456. [PubMed]
29. Smith, P.J.; Olson, J.A.; Constable, D.; Schwartz, J.; Proffitt, R.T.; Adler-Moore, J.P. Effects of dosing regimen on accumulation, retention and prophylactic efficacy of liposomal amphotericin B. *J. Antimicrob. Chemother.* **2007**, *59*, 941–951. [CrossRef] [PubMed]
30. Serrano, D.R.; Hernández, L.; Fleire, L.; González-Alvarez, I.; Montoya, A.; Ballesteros, M.P.; Dea-Ayuela, M.A.; Miró, G.; Bolás-Fernández, F.; Torrado, J.J. Hemolytic and pharmacokinetic studies of liposomal and particulate amphotericin B formulations. *Int. J. Pharm.* **2013**, *447*, 38–46. [CrossRef] [PubMed]
31. Genser, B.; Cooper, P.J.; Yazdanbakhsh, M.; Barreto, M.L.; Rodrigues, L.C. A guide to modern statistical analysis of immunological data. *BMC Immunol.* **2007**, *8*, 27. [CrossRef] [PubMed]
32. Perlin, D.S.; Rautemaa-Richardson, R.; Alastruey-Izquierdo, A. The global problem of antifungal resistance: Prevalence, mechanisms, and management. *Lancet Infect. Dis.* **2017**, *17*, e383–e392. [CrossRef]
33. Chowdhary, A.; Kathuria, S.; Xu, J.; Meis, J.F. Emergence of azole-resistant aspergillus fumigatus strains due to agricultural azole use creates an increasing threat to human health. *PLoS Pathog.* **2013**, *9*, e1003633. [CrossRef]
34. Ignacio, R.M.; Kim, C.-S.; Kim, Y.-D.; Lee, H.-M.; Qi, X.-F.; Kim, S.-K. Therapeutic effect of Active Hexose-Correlated Compound (AHCC) combined with CpG-ODN (oligodeoxynucleotide) in B16 melanoma murine model. *Cytokine* **2015**, *76*, 131–137. [CrossRef] [PubMed]
35. Segarra, S.; Miró, G.; Montoya, A.; Pardo-Marín, L.; Teichenné, J.; Ferrer, L.; Cerón, J.J. Prevention of disease progression in Leishmania infantum-infected dogs with dietary nucleotides and active hexose correlated compound. *Parasites Vectors* **2018**, *11*, 103. [CrossRef] [PubMed]
36. Nogusa, S.; Gerbino, J.; Ritz, B.W. Low-dose supplementation with active hexose correlated compound improves the immune response to acute influenza infection in C57BL/6 mice. *Nutr. Res.* **2009**, *29*, 139–143. [CrossRef]
37. Belay, T.; Fu, C.-L.; Woart, A. Active Hexose Correlated Compound Activates Immune Function to Decrease *Chlamydia trachomatis* Shedding in a Murine Stress Model. *J. Nutr. Med. Diet Care* **2015**, *1*, JNMDC-1-006. [PubMed]

38. Wang, S.; Welte, T.; Fang, H.; Chang, G.-J.J.; Born, W.K.; O'Brien, R.L.; Sun, B.; Fujii, H.; Kosuna, K.; Wang, T. Oral Administration of Active Hexose Correlated Compound Enhances Host Resistance to West Nile Encephalitis in Mice. *J. Nutr.* **2009**, *139*, 598–602. [CrossRef]
39. Ritz, B.W. Supplementation with active hexose correlated compound increases survival following infectious challenge in mice. *Nutr. Rev.* **2008**, *66*, 526–531. [CrossRef]
40. Olson, J.A.; George, A.; Constable, D.; Smith, P.; Proffitt, R.T.; Adler-Moore, J.P. Liposomal amphotericin B and echinocandins as monotherapy or sequential or concomitant therapy in murine disseminated and pulmonary Aspergillus fumigatus infections. *Antimicrob. Agents Chemother.* **2010**, *54*, 3884–3894. [CrossRef]
41. Lewis, R.E.; Albert, N.D.; Kontoyiannis, D.P. Efficacy of Single-Dose Liposomal Amphotericin B or Micafungin Prophylaxis in a Neutropenic Murine Model of Invasive Pulmonary Aspergillosis. *Antimicrob. Agents Chemother.* **2008**, *52*, 4178–4180. [CrossRef] [PubMed]
42. Vetvicka, V.; Vetvickova, J. Immune-enhancing effects of Maitake (*Grifola frondosa*) and Shiitake (*Lentinula edodes*) extracts. *Ann. Transl. Med.* **2014**, *2*, 14. [CrossRef] [PubMed]
43. Daddaoua, A.; Martínez-Plata, E.; Ortega-González, M.; Ocón, B.; Aranda, C.J.; Zarzuelo, A.; Suárez, M.D.; de Medina, F.S.; Martínez-Augustin, O. The nutritional supplement Active Hexose Correlated Compound (AHCC) has direct immunomodulatory actions on intestinal epithelial cells and macrophages involving TLR/MyD88 and NF-κB/MAPK activation. *Food Chem.* **2013**, *136*, 1288–1295. [CrossRef] [PubMed]
44. Yin, Z.; Fujii, H.; Walshe, T. Effects of active hexose correlated compound on frequency of $CD4^+$ and $CD8^+$ T cells producing interferon-$\gamma$ and/or tumor necrosis factor–$\alpha$ in healthy adults. *Hum. Immunol.* **2010**, *71*, 1187–1190. [CrossRef] [PubMed]
45. Thakur, R.; Anand, R.; Tiwari, S.; Singh, A.P.; Tiwary, B.N.; Shankar, J. Cytokines induce effector T-helper cells during invasive aspergillosis; what we have learned about T-helper cells? *Front. Microbiol.* **2015**, *6*, 429. [CrossRef] [PubMed]
46. Dewi, I.M.W.; van de Veerdonk, F.L.; Gresnigt, M.S. The Multifaceted Role of T-Helper Responses in Host Defense against Aspergillus fumigatus. *J. Fungi* **2017**, *3*, 55. [CrossRef] [PubMed]
47. Lee, W.-W.; Lee, N.; Fujii, H.; Kang, I. Active Hexose Correlated Compound promotes T helper (Th) 17 and 1 cell responses via inducing IL-1β production from monocytes in humans. *Cell. Immunol.* **2012**, *275*, 19–23. [CrossRef] [PubMed]
48. Heinekamp, T.; Schmidt, H.; Lapp, K.; Pähtz, V.; Shopova, I.; Köster-Eiserfunke, N.; Krüger, T.; Kniemeyer, O.; Brakhage, A.A. Interference of Aspergillus fumigatus with the immune response. *Semin. Immunopathol.* **2015**, *37*, 141–152. [CrossRef] [PubMed]
49. Goodyear, A.; Jones, A.; Troyer, R.; Bielefeldt-Ohmann, H.; Dow, S. Critical Protective Role for MCP-1 in Pneumonic Burkholderia mallei Infection. *J. Immunol.* **2010**, *184*, 1445–1454. [CrossRef] [PubMed]
50. Ellis, M.; al-Ramadi, B.; Hedström, U.; Alizadeh, H.; Shammas, V.; Kristensen, J. Invasive fungal infections are associated with severe depletion of circulating RANTES. *J. Med. Microbiol.* **2005**, *54*, 1017–1022. [CrossRef] [PubMed]
51. Osherov, N.; Ben-Ami, R. Modulation of Host Angiogenesis as a Microbial Survival Strategy and Therapeutic Target. *PLoS Pathog.* **2016**, *12*, e1005479. [CrossRef]
52. Antachopoulos, C.; Roilides, E. Cytokines and fungal infections. *Br. J. Haematol.* **2005**, *129*, 583–596. [CrossRef] [PubMed]
53. Deepe, G.S. Tumor necrosis factor-alpha and host resistance to the pathogenic fungus, Histoplasma capsulatum. *J. Investig. Dermatol. Symp. Proc.* **2007**, *12*, 34–37. [CrossRef] [PubMed]

*pharmaceutics*

MDPI

*Article*

# Ground Calcium Carbonate as a Low Cost and Biosafety Excipient for Solubility and Dissolution Improvement of Praziquantel

**Ana Borrego-Sánchez [1,2,*], Rita Sánchez-Espejo [1], Beatrice Albertini [3], Nadia Passerini [3], Pilar Cerezo [2], César Viseras [1,2] and C. Ignacio Sainz-Díaz [1]**

1. Instituto Andaluz de Ciencias de la Tierra (CSIC-University of Granada), Av. de las Palmeras 4, 18100 Granada, Spain; ritaespejo@hotmail.com (R.S.-E.); cviseras@ugr.es (C.V.); ignacio.sainz@iact.ugr-csic.es (C.I.S.-D.)
2. Department of Pharmacy and Pharmaceutical Technology, Faculty of Pharmacy, University of Granada, Campus de Cartuja s/n, 18071 Granada, Spain; mcerezo@ugr.es
3. Department of Pharmacy and BioTechnology, University of Bologna, Via S. Donato 19/2, 40127 Bologna, Italy; beatrice.albertini@unibo.it (B.A.); nadia.passerini@unibo.it (N.P.)

* Correspondence: anaborrego@iact.ugr-csic.es; Tel.: +34-958230000

Received: 6 September 2019; Accepted: 12 October 2019; Published: 14 October 2019

**Abstract:** Calcium carbonate is an abundant mineral with several advantages to be a successful carrier to improve oral bioavailability of poorly water-soluble drugs, such as praziquantel. Praziquantel is an antiparasitic drug classified in group II of the Biopharmaceutical Classification System hence characterized by high-permeability and low-solubility. Therefore, the dissolution rate is the limiting factor for the gastrointestinal absorption that contributes to the low bioavailability. Consequently, the therapeutic dose of the praziquantel must be high and big tablets and capsules are required, which are difficult to swallow, especially for pediatric and elderly patients. Mixtures of praziquantel and calcium carbonate using solid-solid physical mixtures and solid dispersions were prepared and characterized using several techniques (X-ray diffraction differential scanning calorimetry, thermogravimetric analysis, scanning electron microscopy, laser diffraction, Fourier transform infrared and Raman spectroscopies). Solubility of these formulations evidenced that the solubility of praziquantel-calcium carbonate interaction product increased in physiological media. *In vitro* dissolution tests showed that the interaction product increased the dissolution rate of the drug in acidic medium. Theoretical models were studied to understand this experimental behavior. Cytotoxicity and cell cycle studies were performed, showing that praziquantel-calcium carbonate physical mixture and interaction product were biocompatible with the HTC116 cells, because it did not produce a decrease in cell viability or alterations in the cell cycle.

**Keywords:** praziquantel; calcium carbonate; schistosomiasis; bioavailability; solubility; cytotoxicity

## 1. Introduction

Calcium carbonate, $CaCO_3$, being a low cost material, with high surface area, excellent safety, biocompatibility and biodegradability, is a well-documented excipient in pharmaceutical solid dosage forms, mainly used as diluent [1]. Recently, it has also been demonstrated that calcium carbonate can successfully act as hydrophilic porous carrier to improve the oral bioavailability of low water soluble drugs [2–5]. Functionalization with hydroxypropyl-β-cyclodextrin [6], enzymatic macromolecules [7], or incorporation in polymeric hydrogels [8–11] can further enhance the ability of calcium carbonate to improve oral delivery of several biomolecules, including proteins [12] and anti-cancer drugs [13]. As recently reviewed, most of the successfully designed calcium carbonate carriers have been prepared by emulsion techniques or chemical precipitation of the carbonate micro/nanoparticles [14]. However, precipitation of calcium carbonate micro/nanoparticles is difficult to reproduce and scale up of the procedure is most of times challenging, avoiding the clinical use of these carriers. Moreover, detailed understanding of biosafety and *in vivo* degradation of the new calcium carbonate particles would require preclinical studies. An interesting alternative is to overcome these challenges by using normalized pharmaceutical excipients grades of calcium carbonate as drug delivery carriers [1].

Praziquantel (PZQ) is the drug of choice in the treatment of schistosomiasis [15], being included in the WHO Model List of Essential Drug for the treatment of adults and children [16]. Schistosomiasis affects approximately 210 million people, causing 200,000 deaths every year. Moreover, it is widely extended, mainly in 78 developing countries in the tropics and subtropics, although at least 92% of people who need treatment for schistosomiasis live in Africa and, is the second of the most prevalent disease (after malaria) affecting African children. PZQ is classified in group II of the Biopharmaceutical Classification System (BCS) and hence characterized by high permeability and low solubility [17]. Therefore, the dissolution rate is the limiting factor for the gastrointestinal absorption that contributes to the low oral bioavailability. Moreover, absorbed amounts of PZQ suffer an extensive first-pass metabolism [18], leading to administration of high-dosed dosage forms, which are difficult to swallow, especially for pediatric patients [19].

Enhancing the water solubility and the dissolution rate is critical to increase the oral bioavailability of PZQ. Several studies and strategies have been carried out to improve PZQ water solubility, such as incorporation into liposome vectors [20], preparation of solid dispersions of the drug with different excipients such as clay minerals [21], β-cyclodextrins [22–24], polyvinyl-pyrrolidone [25–27], polyethylene glycols [28–30], sodium starch glycolate [31] and preparation of dispersible granules [14]. In addition, other techniques have been used to increase the solubility of PZQ such as co-grinding with several excipients [32], melt granulation and ultrasonic spray congealing [27], and the transition in a new crystalline polymorph by milling [19]. More recently, our previous paper evidenced conformational changes of PZQ in association with calcium carbonate after a solvent evaporation process [33]. This interaction product revealed the formation of modifications of the pristine racemic drug attributable to either polymorph B [19] or disordered pseudoracemate solid phases [33].

Given these premises, the first aim of this work was to verify if PZQ solid state modification within calcium carbonate solid dispersion is due to the solvent evaporation process or to drug-carrier interactions that would happen during a mere blending procedure. Thus, the physico-chemical properties of the physical mixture were compared to those of the interaction product. Further, we explored the possibilities of calcium carbonate/PZQ systems for the improvement of biopharmaceutical properties of the drug. In particular, solubility and dissolution profiles of PZQ from interaction product were compared to the corresponding physical mixture, and the results explained by using theoretical modeling. In addition, we investigated the *in vitro* cytotoxicity and cell cycle studies of calcium carbonate/PZQ systems.

## 2. Materials and Methods

### 2.1. Materials

PZQ drug was kindly donated by Fatro S.p.A. (Bologna, Italy). Ground Calcium Carbonate Calcitec Pure PH V/40S (GCC) was purchased from Mineraria Sacilese (Sacile, Italy). This kind of calcium carbonate is used in the pharmaceutical industry and was accepted in the latest edition of the European and United States Pharmacopoeias. The GCC finds also employment in the food industry and in industries where is requested a low content of heavy metals. Ethanol of 96% of purity was used as solvent.

### 2.2. *Preparation of the PZQ and GCC Physical Mixture*

Physical mixtures (PM) of PZQ and GCC (1:5 *w/w*) were prepared blending both solids in an agate mortar of 100 mm of diameter at room temperature for 5 min.

### 2.3. *Preparation of the PZQ and GCC Interaction Product (IP)*

GCC was dispersed in 1 L of ethanolic solution of PZQ under magnetic stirring at room temperature for 24 h, so drug/GCC ratio was 1:5 *w/w* in order to ensure the complete interaction between PZQ and GCC. After 24 h, the solvent was evaporated with rotary evaporator (Rotary evaporator Buchi® R II, Flawil, Switzerland) at 40 °C and reduced pressure. The solid residue was dried in a desiccator and then it was pulverized.

### 2.4. *Solid State Characterization of Calcium Carbonate/Praziquantel Systems*

#### 2.4.1. X-ray Diffraction (XRD)

An X-Pert Pro® diffractometer (Marvel Panalytical, Madrid, Spain) with the CuK$\alpha$ radiation was used for performing powder X-ray diffraction. The powder samples were scanned in the range of 4°–70° of the 2$\theta$ angle, steps were of 0.008 of 2$\theta$ and the counting time was of 10.16 sec/step. The diffraction results were analyzed with the XPOWDER® software version 2004 [34].

#### 2.4.2. Thermal Analysis

Differential scanning calorimetric analysis (DSC) and thermogravimetric analysis (TGA) were performed with a mod. TGA/DSC1 calorimeter (Mettler Toledo, Barcelona, Spain) equipped with a sensor and FRS5 microbalance (precision 0.1 μg) and FP89 software package. Samples were heated in air atmosphere at 5 °C/min in the in the 30–200 °C temperature range for DSC and 30–420 °C temperature range for TGA.

#### 2.4.3. Scanning Electron Microscope (SEM)

Microphotographs of the samples were performed using a Hitachi S-510 scanning electron microscope (voltage 25 kV, secondary electron images) (Hitachi Scientific Instruments Ltd., Tokyo, Japan). The samples were mounted on adhesive carbon paper, fixed with colloidal gold and metallized with gold in two orientations (20–30°). The images were captured digitally using the program attached to the microscope (ScanVision, Version 1.2).

#### 2.4.4. Particle Size Analysis

The particle size distribution of the solid sample materials suspended in liquid medium was analyzed with Laser Light Diffraction technology. The equipment used was a Mastersizer 2000LF from Malvern Panalytical Instruments (Madrid, Spain) consisting of HYDRO MU Malvern manual liquid sample dispersion unit and Malvern HYDRO 2000Up minimum volume liquid sample dispersion unit.

#### 2.4.5. Fourier Transform Infrared and Raman Spectroscopies

Fourier Transform Infrared Spectroscopy (FTIR) spectra were recorded with a 6200 spectrophotometer (JASCO, Pfungstadt, Germany) in the range 4000–600 cm$^{-1}$ with a 0.5 cm$^{-1}$ resolution and a well-plate sampler along with the Spectra Manager II software.

Raman spectra were recorded using a JASCO NRS-5100 Micro-Raman dispersive spectrophotometer in the range 3500–800 cm$^{-1}$ with a 6.48 cm$^{-1}$ resolution, with laser light source VIS-NIR with red diode at 785 nm with 500 mW of power (Torsana Starbright) refrigerated by air and the KnowItAll JASCO for Raman software.

## 2.5. Solubility Studies

The solubility of PZQ as well as the physical mixture PM PZQ-GCC and the interaction product IP PZQ-GCC were studied separately in two media: an acidic medium of HCl 0.001 M and a simulated intestinal fluid (SIF) without enzymes with a buffer at pH 6.8. The solubility of PZQ was calculated by placing a supersaturated solution of the pristine drug, specifically 30 mg in 10 mL in each of the media. The supersaturated solution was stirred in a thermostatic bath for 72 h at 37 °C. After 72 h, it was centrifuged, and the supernatant was filtered and measured on high-performance liquid chromatography (HPLC). Obtaining by means of the HPLC the amount of dissolved drug that corresponds to its solubility. This experiment was repeated 6 times. In the same way this procedure was applied for the PM and the IP PZQ-GCC.

## 2.6. Dissolution Studies

PZQ-GCC physical mixture or PZQ-GCC interaction product (210 mg) were encapsulated in double zero (00) gelatin capsules, corresponding to 35 mg of PZQ and 175 mg of GCC in each capsule. As well as, 35 mg of PZQ were encapsulated as reference. The obtained capsules were subjected to sink conditions dissolution tests using the official Pharmacopoeia USP apparatus 2 for the dissolution test of oral solid dosage forms (Sotax AT7, S). This apparatus is equipped with a rotation system type palettes and sinkers, piston pump for the automatic sampling at scheduled times and collector of fractions. The measurements were performed at 37 °C and 150 rpm in 1 L of medium. Two separate dissolution media were studied: an acidic medium of HCl 0.001 M (simulated stomach) and a SIF medium without enzymes with a buffer at pH 6.8 (simulated intestine). Aliquots of 5 mL were collected from the dissolution test, filtered through 0.45 μm Millipore® (S) membranes and analyzed by HPLC for drug content. Volumes of 5 mL of fresh dissolution medium were replaced after each sampling to maintain the volume constant. At least three replicates for each sample were assayed.

## 2.7. HPLC Analysis

Drug analysis was performed using a 1260 Infinity II Agilent HPLC system (Santa Clara, CA, USA) equipped with quaternary pump, autosampler, column oven and UV-VIS diode-array spectrophotometer. The stationary phase was a Kromasil® C18 column, 5 μm, 250 × 4.6 mm (Teknokroma, Barcelona, Spain) and the mobile phase was a mixture of $H_2O$ and $CH_3CN$ (35:65 *v/v*). The flow rate was set at 0.8 mL/min with an injection volume of 10 μL. A spectrophotometer detector at a 225 nm wavelength was used and the run time for each analysis was 5 min. Data were recorded and analyzed by using software LC Open LAB HPLC 1260 (Agilent). The response of the analytical method was linear in the concentration range 5–100 mg/L of PZQ in both media, resulting in correlation coefficients of 1 (both in HCl 0.001 M and SIF).

## 2.8. Computational Methods

The molecular model of PZQ was extracted from previous calculations [35] and the rest of components were placed by hand. The Compass force field (FF) [36] based on empirical interatomic potentials was used within the Discover program of the Materials Studio package [37]. This FF was used previously to describe PZQ molecules and crystal structure with satisfactory results [35,38]. An atomic interactions cut-off of 18.5 Å was used for calculating Van der Waals and Coulomb interactions.

## 2.9. Cell Culture

Cell viability tests and study the cell cycle profiles were performed for observing a possible production of cell death in a tumor line of cells derived from colorectal carcinoma, called HCT116. This tumor line of HCT116 colorectal carcinoma cells were cultured in Dulbecco's Modified Eagle's Medium (DMEM) (Gibco Invitrogen, Dublin, Ireland) supplemented 10% with decomplemented Heat-Inactivated Fetal Bovine Serum (FBS) (Gibco Invitrogen), with glutamax (BioWhittaker, Cologne,

Germany) at 1% and with Penicillin/Streptomycin (BioWhittaker) at 1%. The cell culture was kept in an incubator at 37 °C and 5% of $CO_2$.

### 2.10. Cytotoxicity Studies

Firstly, PZQ and GCC samples were prepared relying on the administered drug amount in animals as antiparasitic. In such way, the amount of the mineral was five times more than that of PZQ. To do this, we prepared an intermediate dilution in dimethylsulfoxide (DMSO) at a concentration of 100 mM and then another dilution at 10 mM. The physical mixture and the interaction product PZQ-GCC were studied using the same procedure.

To perform the proliferation assays by Alamar Blue, about 10,000–20,000 cells/well of HCT116 were seeded in 96-well plates, in a final volume of 200 μL of DMEM medium using several concentrations (100 μM, 20 μM, 4 μM, 800 nM, 160 nM, 32 nM and 6.4 nM) for each sample studied (PZQ, GCC, PM and IP PZQ-GCC). Cells were incubated at 37 °C, 5% $CO_2$ for 48 h. After the incubation time, 10 μL per well of the PrestoBlue cell Viability Reagent (Invitrogen, Thermo Fisher Scientific, Carlsbad, CA, USA) was added, incubating for 15 min. After this period, fluorescence was measured at 535–90 nm in a Tecan reader (Männedorf, Switzerland). This reagent is based on resazurin compound, which works as an indicator of viability when it is reduced by living cells and presents a colour change from blue (dead cells) to pink (living cells).

### 2.11. Cell Cycle Studies

The cells were cultured in 24 well plates at a concentration of 250,000 cells/well in 500 μL and treated with the previously selected PZQ, GCC, PM and IP PZQ-GCC samples, at increasing concentrations of 0.8, 4, 20 and 100 μM for 48 h. Dead cells (apoptotic and necrotic) were detected based on staining with propidium iodide following the protocol described [39]. Briefly, the cells after the corresponding treatments were collected and washed with 2 mL of phosphate buffered saline (PBS) at 4 °C and fixed with 100 μL of PBS and 900 μL of 70% ethanol on ice for 5 min. After washing with PBS, they were resuspended in 250 μL of PBS and another 250 μL of a DNA extraction solution (0.2 M $Na_2HPO_4$, 0.1M C6H8O7, pH 7.8) and incubated at 37 °C for 10 min. The supernatant was removed and 200 μL of the staining solution was added (8 μL propidium iodide (1 mg/mL) and 2 μL RNAse 100 (μg/mL)), incubating the samples for 10 min at 37 °C in the dark. Fluorescence was measured in the FL2 detector of the FACScalibur cytometer (Becton Dickinson & Co., Franklin Lakes, NJ, USA) and the analysis of the sub-G1 population (population of dead cells: necrotic and apoptotic) was done using the BD CellQuest software v1.0.2 (BD, Biosciences).

## 3. Results and Discussion

### 3.1. Solid State Characterization of Calcium Carbonate/Praziquantel Systems

PZQ and GCC (1:5) solid dispersions obtained by solvent evaporation evidenced conformational changes of the pristine drug solid state [33]. Recently, a number of papers [19,32,40] have elucidated that the mechanochemical activation of PZQ, via neat grinding or comilling PZQ with different polymers (Povidone, Copovidone, Crospovidone and Sodium Starch Glycolate) enabled the transformation of the original PZQ polymorphic form into a new polymorphic variety of racemic PZQ or into a drug amorphous state.

In this study, the properties of PZQ-GCC physical mixture (produced by blending) were compared to those of the PZQ-GCC interaction product (obtained by solvent evaporation). A solid state characterization of the samples studied was performed by X-ray diffraction (Figure S1), differential scanning calorimetric analysis (Figure S2), thermogravimetric analysis (Figure S3), scanning electron microscope (Figure S4), Fourier Transform Infrared and Raman Spectroscopies (Figure S5) and particle size analysis. The results demonstrated that PZQ-GCC physical mixture did not show any chemical-physical change with respect to the pure drug. On the contrary, in the PZQ-GCC interaction product multiple changes were observed, as already described in a previous work [33]. The detailed results and analyses of the solid state characterization of the samples studied (PZQ, GCC, PM and IP) are provided in the Supplementary Material.

### 3.2. Solubility Studies

The solubility tests showed that the IP PZQ-GCC increased the solubility of the pure drug from 0.50 mg/mL to 1.42 mg/mL in an acid aqueous medium with 0.001 M HCl (pH = 3), due to its interaction with the GCC and the solubility of the GCC in acidic medium. On the contrary, the PM PZQ-GCC showed a small and unimportant increase in the solubility of the drug (Table 1). The results in a SIF medium (pH = 6.8) revealed that the IP PZQ-GCC increased the solubility of PZQ from 0.45 mg/mL to 0.73 mg/mL; while the PM showed a slight increase in the solubility of PZQ from 0.45 mg/mL to 0.50 mg/mL, which was not considered notable. Therefore, the solubility of the IP PZQ-GCC improves in acid and SIF media with respect to the PZQ and PM. This increase is much higher at pH 3 (Table 1).

**Table 1.** Solubility values of praziquantel (PZQ), physical mixture (PM) of PZQ-GCC and interaction product (IP) of PZQ-GCC in acid and simulated intestinal fluid (SIF) media (mean values ± standard deviation; $n = 0.07$).

| Sample | HCl 0.001 M Medium, pH = 3 | SIF Medium, pH = 6.8 |
|---|---|---|
| PZQ | 0.50 mg/mL | 0.45 mg/mL |
| PM PZQ-GCC | 0.52 mg/mL | 0.50 mg/mL |
| IP PZQ-GCC | 1.42 mg/mL | 0.73 mg/mL |

### 3.3. Dissolution Studies

In order to explore the *in vitro* bioavailability of PZQ, dissolution tests in both media simulating the gastric and the intestinal fluids, were performed for the pristine drug and the combinations of PZQ with GCC (IP and PM) (Figure 1).

PZQ-GCC physical mixture presented a similar behavior to the pristine PZQ in both dissolution media (Figure 1a,b), while the IP showed a completely different behavior at pH 3 with respect to pH 6.8. Interaction product PZQ-GCC showed an increase in the dissolution rate in acid medium (Figure 1a) and a strong decrease in SIF (Figure 1b). These results were concordant with solubility tests results, where the solubility of IP PZQ-GCC is higher in acid medium than in SIF. This behaviour may be correlated to the different solubility of GCC at acidic and neutral pH and to structural changes observed in IP: as GCC is soluble at low pH, its interaction at molecular level with water and PZQ favoured the dissolution and solubility of the drug. Likewise, the PZQ-GCC interaction at pH 6.8, where the carbonate is completely insoluble, negatively affected the solubilisation of the PZQ molecules.

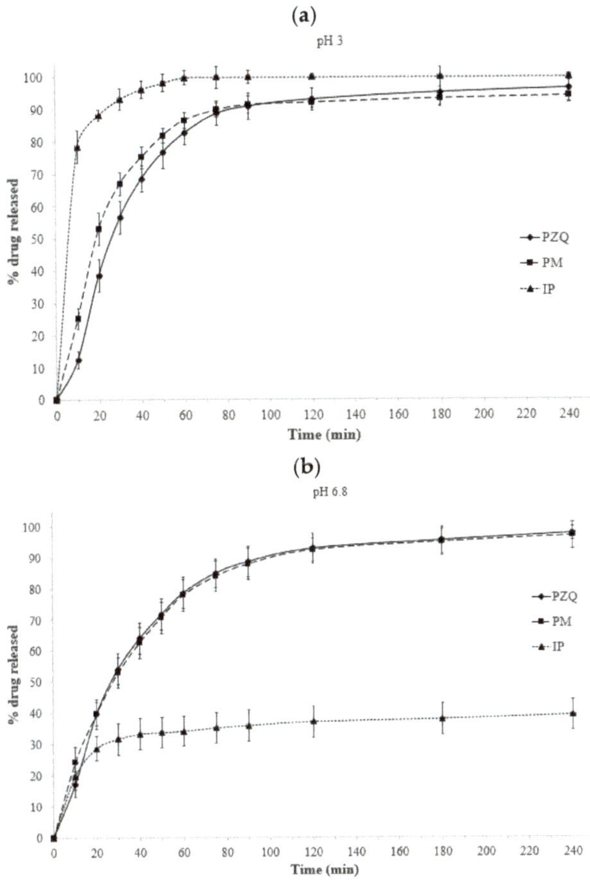


**Figure 1.** Drug release (% *w/w*) profiles from the PZQ and the PM in 0.001 M HCl at pH = 3 (**a**) and in a SIF medium at pH = 6.8 (**b**) in sink conditions; (mean values ± standard deviation; $n = 5$).

### 3.4. Modeling Approach

In order to explain the dissolution behavior of the PZQ-GCC systems, small models of PZQ molecules with the GCC and water molecules were created (Figure 2). In the PM, the carbonyl groups of PZQ are at the same side of the molecule as a *syn* conformer, whereas in IP the proportion of the *anti* conformer is higher [38]. The geometry of these models was optimized with Compass FF calculations, using the SPC water model (Figure 2c,d). In both models the carbonate anion is coordinated with the $Ca^{2+}$ cation $d$(Ca … OCO) = 1.88–1.91 Å and at least one water molecule is between the $Ca^{2+}$ cation and a carbonyl group forming a hydrogen bond with the corresponding carbonyl group $d$(HOH … O=C) = 1.84–1.88 Å. In the *syn* conformer, both water molecules are coordinating the $Ca^{2+}$ cation and the water H atoms $d$(HOH … Ca) = 2.27 Å (Figure 2c).

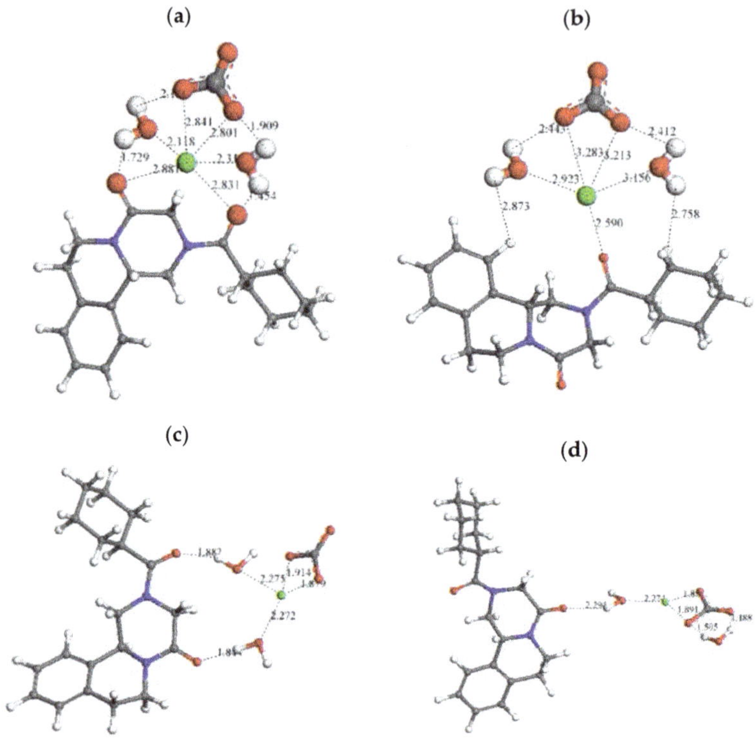


**Figure 2.** Possible complexes of PZQ molecule conformer *syn* (**a**) and *anti* (**b**) with hydrated $Ca^{2+}$ carbonates clusters, and optimized complexes of *syn* (**c**) and *anti* (**d**) PZQ (non-bonding distances are included in Å). The C, Ca, H, N, and O atoms are in grey, green, white, blue, and red color.

In the *anti* conformer, one water molecule is coordinating the $Ca^{2+}$ cation, d(O ... Ca) = 2.27 Å, and one carbonyl O atom d(CO ... H) = 2.22 Å leaving the another carbonyl group free in the opposite side of the molecule to form an additional carbonate complex. The conformer *syn* can form only one complex with the hydrated $Ca^{2+}$ cation (Figure 2). This could explain the higher dissolution process of the *anti* form of PZQ and hence the higher solubility of IP PZQ-GCC.

### 3.5. Cytotoxicity Studies

*In vitro* cytotoxicity tests in the HCT116 cell line showed that all the PZQ and GCC pure concentrations (6.4 nM–100 μM) tested provided viability values slightly lower than the untreated cells (control). Therefore, PZQ and GCC pure samples can be considered biocompatible toward HCT116 cell lines (Figure 3). In the IP and PM PZQ-GCC products similar results were found, whose biocompatibility across the range of concentrations was evaluated and a trend similar to the control was observed and slightly positive with respect to the pure samples (PZQ and GCC) in cell growth. Therefore, it is notable that IP and PM PZQ-GCC products have a biocompatible behaviour, so the interaction between PZQ and GCC had a positive effect in the cell viability (Figure 3).

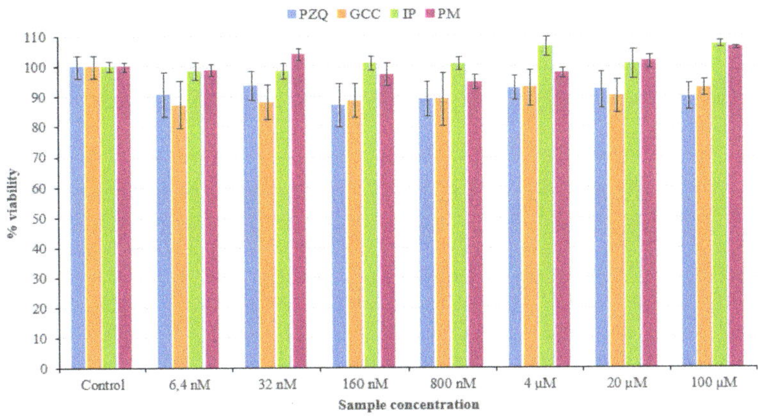


**Figure 3.** Cell viability from the studied samples after 48 h of treatment. (Control: untreated cells in complete medium; mean values ± standard error; $n = 8$).

### 3.6. Cell Cycle Studies

A study of the cell cycle by means of propidium iodide was carried out to study the cell cycle corresponding to the Sub-G1 phase of the cells in contact with our samples (Figure 4). Once observed through cytotoxicity studies that the samples tested did not cause cell death, a cell cycle study was carried out to check whether the compounds affected any phase of the cell cycle despite not affecting proliferation. In the upper part of the figure the control tests can be observed: an example of the cell cycle of healthy untreated cells, an example of cells treated with DMSO, where it is observed that DMSO does not affect the cell and also an example of cells treated with etoposide, an antineoplastic drug that damages cells. In this last example, the cell cycle of the control cells is severely affected and induces death to 64.3% of the cell population.

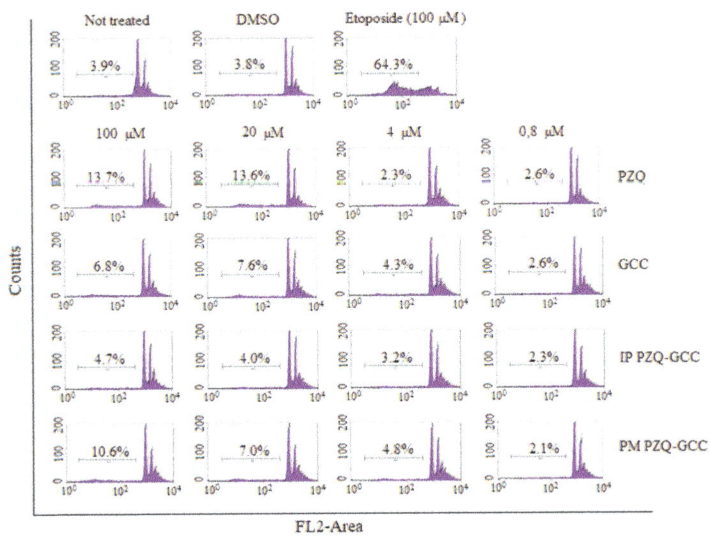


**Figure 4.** Cell cycle of the HCT116 line treated with the sample studies. The percentages indicate the number of cells in Sub-G1 (apoptotic or necrotic).

The studied samples showed that they do not affect in any case the cellular cycle of the cells, only in some cases at high concentrations can the cell death increase slightly, although not in a noteworthy way (Figure 4).

## 4. Conclusions

Interaction products obtained with PZQ and GCC increased the solubility of PZQ in physiological media, particularly in acid medium. *In vitro* dissolution tests evidenced that the interaction product increased the dissolution rate of the drug in acidic medium. *In vitro* cytotoxicity and cell cycle studies were performed, showing neither the PZQ-GCC physical mixture nor the interaction product produced cellular damage in any of the studied concentrations. Therefore, the samples studied did not produce cell death or alteration in the cell cycle and were biocompatible with the HTC116 cells.

As a general conclusion, the use of low cost GCCimprove solubility and dissolution rates of PZQ, without producing cytotoxicity or alterations in the cell cycle of HTC116 cells. This allows exploring the use of GCC as an interesting technological strategy for a drug administration in a more effective biopharmaceutical and clinical way ofPZQ. Also providing the great advantage that the design and development of the PZQ-GCC new formulations would not increase the final cost of the drug, overcoming a great challenge due to the population which the treatment is destined and the high prevalent of the neglected tropical disease.

**Supplementary Materials:** The following are available online at http://www.mdpi.com/1999-4923/11/10/533/s1, Section 1: Solid State Characterization of Calcium Carbonate/Praziquantel Systems. Figure S1: Powder X-ray diffraction (XRD) patterns of praziquantel (PZQ), ground calcium carbonate (GCC), physical mixture (PM) of PZQ-GCC and interaction product (IP) of PZQ-GCC. Figure S2: DSC curves of PZQ, GCC, PM and IP. Figure S3: TGA profiles of the GCC, IP, PM, and PZQ solids, Figure S4: SEM micropictures of GCC, pristine PZQ, PM and IP. Figure S5: FTIR (a), and Raman (b) spectra of PZQ, GCC, IP and PM.

**Author Contributions:** Conceptualization, B.A., N.P., C.V. and C.I.S.-D.; methodology, A.B.-S., C.V. and C.I.S.-D.; software, A.B.-S. and C.I.S.-D.; formal analysis, A.B.-S. and R.S.-E.; investigation, A.B.-S. and R.S.-E.; resources, P.C., C.V. and C.I.S.-D.; writing—original draft preparation, A.B.-S.; writing—review and editing, A.B.-S., B.A., N.P., P.C., R.S.-E., C.V. and C.I.S.-D.; supervision, A.B.-S., C.V. and C.I.S.-D.; project administration, A.B.-S.; funding acquisition, P.C., C.V. and C.I.S.-D.

**Acknowledgments:** The authors are thankful to Fatro S.p.A. for donating the praziquantel drug and to Xtrem Biotech, specially Ignacio Molina and Sara Torres, for their contributions with the cytotoxicity and cell cycle experiments. We also acknowledge for financial support the MINECO, for projects FIS2016-77692-C2-2-P and CGL2016-80833-R, and the Andalusian government, for project RNM1897.

**Conflicts of Interest:** The authors declare no conflict of interest.

## References

1. Rowe, R.C.; Sheskey, P.J.; Weller, P.J. Calcium Carbonate. In *Handbook of Pharmaceutical Excipients*, 6th ed.; Pharmaceutical Press: London, UK, 2009.
2. Ueno, Y.; Futagawa, H.; Takagi, Y.; Ueno, A.; Mizushima, Y. Drug-incorporating calcium carbonate nanoparticles for a new delivery system. *J. Control. Release* **2005**, *103*, 93–98. [CrossRef] [PubMed]
3. Preisig, D.; Haid, D.; Varum, F.J.; Bravo, R.; Alles, R.; Huwyler, J.; Puchkov, M. Drug loading into porous calcium carbonate microparticles by solvent evaporation. *Eur. J. Pharm. Biopharm.* **2014**, *87*, 548–558. [CrossRef] [PubMed]
4. Manzine Costa, L.M.; Molina de Olyveira, G.; Salomão, R. Precipitated Calcium Carbonate Nano-Microparticles: Applications in Drug Delivery. *Adv. Tissue Eng. Regen. Med.* **2017**, *3*, 2.
5. Dizaj, S.M.; Barzegar-Jalali, M.; Zarrintan, M.H.; Adibkia, K.; Lotfipour, F. Calcium carbonate nanoparticles as cancer drug delivery system. *Expert Opin. Drug Deliv.* **2015**, *12*, 1649–1660. [CrossRef]
6. Zhang, L.; Zhu, W.; Lin, Q.; Han, J.; Jiang, L.; Zhang, Y. Hydroxypropyl-β-cyclodextrin functionalized calcium carbonate microparticles as a potential carrier for enhancing oral delivery of water-insoluble drugs. *Int. J. Nanomed.* **2015**, *10*, 3291–3302. [CrossRef]

7. Islan, G.A.; Cacicedo, M.L.; Bosio, V.E.; Castro, G.R. Development and characterization of new enzymatic modified hybrid calcium carbonate microparticles to obtain nano-architectured surfaces for enhanced drug loading. *J. Colloid Interface Sci.* **2015**, *439*, 76–87. [CrossRef]
8. Wang, J.; Chen, J.-S.; Zong, J.-Y.; Zhao, D.; Li, F.; Zhuo, R.-X.; Cheng, S.-X. Calcium Carbonate/Carboxymethyl Chitosan Hybrid Microspheres and Nanospheres for Drug Delivery. *J. Phys. Chem. C* **2010**, *114*, 18940–18945. [CrossRef]
9. Wang, C.; Liu, H.; Gao, Q.; Liu, X.; Tong, Z. Alginate-calcium carbonate porous microparticle hybrid hydrogels with versatile drug loading capabilities and variable mechanical strengths. *Carbohydr. Polym.* **2008**, *71*, 476–480. [CrossRef]
10. Wang, J.; Kong, Y.; Liu, F.; Shou, D.; Tao, Y.; Qin, Y. Construction of pH-responsive drug delivery platform with calcium carbonate microspheres induced by chitosan gels. *Ceram. Int.* **2018**, *44*, 7902–7907. [CrossRef]
11. Vergaro, V.; Papadia, P.; Leporatti, S.; De Pascali, S.A.; Fanizzi, F.P.; Ciccarella, G. Synthesis of biocompatible polymeric nano-capsules based on calcium carbonate: A potential cisplatin delivery system. *J. Inorg. Biochem.* **2015**, *153*, 284–292. [CrossRef]
12. Roth, R.; Schoelkopf, J.; Huwyler, J.; Puchkov, M. Functionalized calcium carbonate microparticles for the delivery of proteins. *Eur. J. Pharm. Biopharm.* **2018**, *122*, 96–103. [CrossRef] [PubMed]
13. Svenskaya, Y.; Parakhonskiy, B.; Haase, A.; Atkin, V.; Lukyanets, E.; Gorin, D.; Antolini, R. Anticancer drug delivery system based on calcium carbonate particles loaded with a photosensitizer. *Biophys. Chem.* **2013**, *182*, 11–15. [CrossRef] [PubMed]
14. Trofimov, A.D.; Ivanova, A.A.; Zyuzin, M.V.; Timin, A.S. Porous Inorganic carriers Based on Silica, Calcium Carbonate and Calcium Phosphate for Controlled/Modulated Drug Delivery: Fresh Outlook and Future Perspectives. *Pharmaceutics* **2018**, *10*, 167. [CrossRef] [PubMed]
15. Trastullo, R.; Dolci, L.S.; Passerini, N.; Albertini, B. Development of flexible and dispersible oral formulations containing praziquantel for potential schistosomiasis treatment of pre-school age children. *Int. J. Pharm.* **2015**, *495*, 536–550. [CrossRef] [PubMed]
16. WHO, World Health Organization. Schistosomiasis. Available online: http://www.who.int/es/news-room/fact-sheets/detail/schistosomiasis (accessed on 12 October 2019).
17. Lindenberg, M.; Kopp, S.; Dressman, J.B. Classification of orally administered drugs on the World Health Organization Model list of Essential Medicines according to the biopharmaceutics classification system. *Eur. J. Pharm. Biopharm.* **2004**, *58*, 265–278. [CrossRef]
18. El-Subbagh, H.I.; Al-Badr, A.A. Praziquantel. In *Analytical Profiles of Drug Substances and Excipients*; Academic Press Inc.: New York, NY, USA, 1998; Volume 25, pp. 463–500.
19. Zanolla, D.; Perissutti, B.; Passerini, N.; Chierotti, M.R.; Hasa, D.; Voinovich, D.; Gigli, L.; Demitri, N.; Geremia, S.; Keiser, J.; et al. A new soluble and bioactive polymorph of praziquantel. *Eur. J. Pharm. Biopharm.* **2018**, *127*, 19–28. [CrossRef]
20. Mourao, S.C.; Costa, P.I.; Salgado, H.R.N.; Gremiao, M.P.D. Improvement of antischistosomal activity of praziquantel by incorporation into phosphatidylcholine-containing liposomes. *Int. J. Pharm.* **2005**, *295*, 157–162. [CrossRef]
21. Borrego-Sánchez, A.; Carazo, E.; Aguzzi, C.; Viseras, C.; Sainz-Díaz, C.I. Biopharmaceutical improvement of praziquantel by interaction with montmorillonite and sepiolite. *Appl. Clay Sci.* **2018**, *160*, 173–179. [CrossRef]
22. Rodrigues, S.G.; Chaves, I.S.; Melo, N.F.S.; de Jesus, M.B.; Fraceto, L.F.; Fernandes, S.A.; de Paula, E.; de Freitas, M.P.; Pinto, L.M.A. Computational analysis and physico-chemical characterization of an inclusion compound between praziquantel and methyl-$\beta$-cyclodextrin for use as an alternative in the treatment of schistosomiasis. *J. Incl. Phenom. Macrocycl. Chem.* **2011**, *70*, 19–28. [CrossRef]
23. Maragos, S.; Archontaki, H.; Macheras, P.; Valsami, G. Effect of Cyclodextrin Complexation on the Aqueous Solubility and Solubility/Dose Ratio of Praziquantel. *AAPS Pharm. Sci. Tech.* **2009**, *10*, 1444. [CrossRef]
24. Becket, G.; Schep, L.J.; Tan, M.Y. Improvement of the in vitro dissolution of praziquantel by complexation with $\alpha$-, $\beta$-, and $\gamma$-cyclodextrins. *Int. J. Pharm.* **1999**, *179*, 65–71. [CrossRef]
25. De la Torre, P.; Torrado, S.; Torrado, S. Preparation, Dissolution and Characterization of Praziquantel Solid Dispersions. *Chem. Pharm. Bull.* **1999**, *47*, 1629–1633. [CrossRef]
26. El-Arini, S.K.; Leuenberger, H. Dissolution properties of praziquantel-PVP systems. *Pharm. Acta Helv.* **1998**, *73*, 89–94. [CrossRef]

27. Costa, E.D.; Priotti, J.; Orlandi, S.; Leonardi, D.; Lamas, M.C.; Nunes, T.G.; Diogo, H.P.; Salomon, C.J.; Ferreira, M.J. Unexpected solvent impact in the crystallinity of praziquantel/poly(vinylpyrrolidone) formulations. A solubility, DSC and solid-state NMR study. *Int. J. Pharm.* **2016**, *511*, 983–993. [CrossRef]
28. Passerini, N.; Albertini, B.; Perissutti, B.; Rodriguez, L. Evaluation of melt granulation and ultrasonic spray congealing as techniques to enhance the dissolution of praziquantel. *Int. J. Pharm.* **2006**, *318*, 92–102. [CrossRef]
29. Liu, Y.; Wang, T.; Ding, W.; Dong, C.; Wang, X.; Chen, J.; Li, Y. Dissolution and oral bioavailability enhancement of praziquantel by solid dispersions. *Drug Deliv. Transl. Res.* **2018**, *8*, 580–590. [CrossRef]
30. Cheng, L.; Lei, L.; Guo, S. In vitro and in vivo evaluation of praziquantel loaded implants based on PEG/PCL blends. *Int. J. Pharm.* **2010**, *387*, 129–138. [CrossRef]
31. Chaud, M.V.; Lima, A.C.; Vila, M.M.D.C.; Paganelli, M.O.; Paula, F.C.; Pedreiro, L.N.; Gremião, M.P.D. Development and evaluation of praziquantel solid dispersions in sodium starch glycolate. *Trop. J. Pharm. Res.* **2013**, *12*, 163–168. [CrossRef]
32. Perissutti, B.; Passerini, N.; Trastullo, R.; Keiser, J.; Zanolla, D.; Zingone, G.; Voinovich, D.; Albertini, B. An explorative analysis of process and formulation variables affecting comilling in a vibrational mill: The case of praziquantel. *Int. J. Pharm.* **2017**, *533*, 402–412. [CrossRef]
33. Borrego-Sánchez, A.; Carazo, E.; Albertini, B.; Passerini, N.; Perissutti, B.; Cerezo, P.; Viseras, C.; Aguzzi, C.; Sainz-Díaz, C.I. Conformational polymorphic changes in the crystal structure of the chiral antiparasitic drug praziquantel and interactions with calcium carbonate. *Eur. J. Pharm. Biopharm.* **2018**, *132*, 180–191. [CrossRef]
34. Martín-Ramos, J.D. XPowder software. 2005. Available online: http://www.xpowder.com (accessed on 12 October 2019).
35. Borrego-Sánchez, A.; Viseras, C.; Aguzzi, C.; Sainz-Díaz, C.I. Molecular and crystal structure of praziquantel. Spectroscopic properties and crystal polymorphism. *Eur. J. Pharm. Sci.* **2016**, *92*, 266–275. [CrossRef] [PubMed]
36. Sun, H. COMPASS: An ab Initio Force-Field Optimized for Condensed-Phase Applications-Overview with Details on Alkane and Benzene Compounds. *J. Phys. Chem. B* **1998**, *102*, 7338–7364. [CrossRef]
37. *Accelrys Materials Studio*, version v 6.0; Accelrys Inc.: San Diego, CA, USA, 2012.
38. Borrego-Sánchez, A.; Hernández-Laguna, A.; Sainz-Díaz, C.I. Molecular modeling and infrared and Raman spectroscopy of the crystal structure of the chiral anti-parasitic drug praziquantel. *J. Mol. Model.* **2017**, *23*, 106. [CrossRef] [PubMed]
39. Gong, J.P.; Traganos, F.; Darzynkiewicz, Z. A selective procedure for DNA extraction from apoptotic cells applicable for gel electrophoresis and flow cytometry. *Anal. Biochem.* **1994**, *218*, 314–319. [CrossRef] [PubMed]
40. Zanolla, D.; Perissutti, B.; Passerini, N.; Invernizzi, S.; Voinovich, D.; Bertoni, S.; Melegari, C.; Millotti, G.; Albertini, B. Milling and comilling Praziquantel at cryogenic and room temperatures: Assessment of the process-induced effects on drug properties. *J. Pharm. Biomed. Anal.* **2018**, *153*, 82–89. [CrossRef] [PubMed]

*pharmaceutics*

MDPI

*Article*

# Resorbable Beads Provide Extended Release of Antifungal Medication: In Vitro and In Vivo Analyses

**Yung-Heng Hsu [1], Huang-Yu Chen [2], Jin-Chung Chen [3], Yi-Hsun Yu [1], Ying-Chao Chou [1], Steve Wen-Neng Ueng [1,*] and Shih-Jung Liu [1,2,*]**

1 Department of Orthopedic Surgery, Chang Gung Memorial Hospital-Linkou, Tao-Yuan 33305, Taiwan; laurencehsu.hsu@gmail.com (Y.-H.H.); alanyu1007@gmail.com (Y.-H.Y.); enjoycu@ms22.hinet.net (Y.-C.C.)
2 Department of Mechanical Engineering, Chang Gung University, Tao-Yuan 33302, Taiwan; kyoiorikyo@gmail.com
3 Graduate Institute of Biomedical Science, Chang Gung University, Tao-Yuan 33302, Taiwan; Jinchen@mail.cgu.edu.tw
* Correspondence: wenneng@adm.cgmh.org.tw (S.W.-N.U.); profsjliu5347@gmail.com (S.-J.L.); Tel.: +886-3-3281200 (S.W.-N.U.); +886-3-2118166 (S.-J.L.)

Received: 15 September 2019; Accepted: 22 October 2019; Published: 24 October 2019

**Abstract:** Fungal osteomyelitis has been difficult to treat, with first-line treatments consisting of implant excision, radical debridement, and local release of high-dose antifungal agents. Locally impregnated antifungal beads are another popular treatment option. This study aimed to develop biodegradable antifungal-agent-loaded Poly(D,L-lactide-*co*-glycolide) (PLGA) beads and evaluate the in vitro/in vivo release patterns of amphotericin B and fluconazole from the beads. Beads of different sizes were formed using a compression-molding method, and their morphology was evaluated via scanning electron microscopy. Intrabead incorporation of antifungal agents was evaluated via Fourier-transform infrared spectroscopy, and in vitro fluconazole liberation curves of PLGA beads were inspected via high-performance liquid chromatography. When we implanted the drug-incorporated beads into the bone cavity of rabbits, we found that a high level of fluconazole (beyond the minimum therapeutic concentration [MTC]) was released for more than 49 d in vivo. Our results indicate that compression-molded PLGA/fluconazole beads have potential applications in treating bone infections.

**Keywords:** fluconazole; orthopedic infection; Poly(D,L-lactide-*co*-glycolide) beads; sustained release

## 1. Introduction

Fungal osteomyelitis is an uncommon disease presenting significant challenges to orthopedic surgeons. The severity of a fungal infection is associated with the immune status of individuals and the fungal species. A large number of fungal infections have been reported in both immunocompromised and immunocompetent individuals [1]. Fungal infections typically result from three routes: direct inoculation, hematogenous spreading, or contiguous spreading [2,3]. In certain conditions, including long-term antibiotic use, infections of *Candida*, *Aspergillus*, and other common fungi are characterized by the formation of a biofilm that resists antifungal treatment, thus further strengthening the infection [4]. Certain fungal infections can exclusively be treated with antifungal agents [5]. However, most patients are difficult to treat via radical debridement. General principles for treatment of both bacterial osteomyelitis and fungal infections include excision of all nonviable tissue, the use of orthopedic hardware, radical debridement, and local drug delivery with an effective concentration of antifungal/antibiotic agents [6,7].

The most common fungi causing bone and joint infections are *Candida* spp., and *Candida albicans* in particular [8]. Fluconazole is a first-generation triazole antifungal agent commonly used to treat

*Candida albicans* infections [9]. Furthermore, amphotericin B sodium deoxycholate effectively inhibits biofilm formation in multiple *Candida* infections [10]. Treatment of *Candida* osteomyelitis involves surgical debridement and long-term administration of antifungal agents. The guidelines of the Infectious Diseases Society of America (IDSA) for the treatment of *Candida* osteomyelitis suggest the administration of fluconazole for 6–12 months [11]. Currently, antifungal-agent-containing beads or polymethyl methacrylate (PMMA) spacers are popular treatment methods for fungal osteomyelitis or periprosthetic joint fungal infections [4]. The beads or spacers delivered to bone tissue display sustained long-term release of antifungal agents at an effective concentration. However, the potential risk of irritation in host tissues combined with the non-biodegradable nature of PMMA may limit application in osteomyelitis treatment, thus warranting surgical excision. Kweon et al. reported that adding 10 g poragen to antifungal-loaded bone cement (ALBC) containing 200 mg amphotericin B decreases the compressive strength of PMMA beads and thus limits their use for implant fixation [12]. Sealy et al. showed poor release dynamics of fluconazole in ALBC [13]. Furthermore, Goss et al. reported that amphotericin B could not be eluted through PMMA bone cement [14]. An ideal drug delivery system should provide adequate antifungal concentrations at the target site, offer a slow and sustained release of an antimicrobial over an extended period, and be biodegradable so that a second operation is not needed.

Biodegradable antifungal-agent-loaded beads possess advantages over conventional PMMA beads in four ways. First, biodegradable beads provide high concentrations of antifungal agents for the extended time needed to completely treat the particular orthopedic infection. Second, variable biodegradability from weeks to months permits various types of infections to be treated. Third, the biodegradable vehicles degrade eventually, and surgical removal of the beads is not required. Fourth, the biodegradable beads dissolve gradually and the soft tissue or bone defect slowly fills with tissue, it is thus not necessary for bone/tissue reconstruction [15].

This current study developed biodegradable antifungal-agent-loaded vehicles for a long-term drug release. We utilized a compression-molding method to fabricate fluconazole-incorporated Poly(D,L-lactide-*co*-glycolide) (PLGA) beads and assessed drug release dynamics in vitro and in vivo. Among the various polymeric materials available to develop local drug release systems, PLGA is promising, owing to its degradable polymer that facilitates long-term drug delivery at high doses to the target region [16,17]. Moreover, this material has been certified for clinical use owing to its nontoxic nature and minimum inflammatory effects.

After molding, bead morphology was evaluated via scanning electron microscopy (SEM), and intrabead antifungal drug incorporation was assessed via Fourier-transform infrared (FTIR) spectroscopy. In vitro fluconazole liberation curves of PLGA beads were inspected via high-performance liquid chromatography (HPLC), and in vivo fluconazole release was investigated by implanting the drug-incorporated beads into the bone cavity of rabbits. We found that a level of fluconazole greater than the minimum therapeutic concentration (MTC) was released for more than 49 d in vivo, indicating that compression-molded PLGA/fluconazole beads are a promising candidate for the treatment of bone infections.

## 2. Materials and Methods

### 2.1. *Fabrication of Poly(D,L-lactide-co-glycolide) PLGA/Amphotericin B and PLGA/Fluconazole Beads*

All materials utilized in this study, including PLGA (50:50), amphotericin B, and fluconazole, were acquired from Sigma-Aldrich (St. Louis, MO, USA).

PLGA/amphotericin B and PLGA/fluconazole beads were prepared using a laboratory-scale compression-molding system equipped with an isothermal oven with a temperature range of 25 °C to 300 °C. Beads of two different polymer:drug ratios (6:1 and 4:1) and three sizes (3, 5, and 8 mm) were fabricated. PLGA, amphotericin B, and fluconazole at predetermined weights were first mixed using a dry mixer and placed in molds (Figure 1) customized for this study. Table 1 lists the composition of beads of different sizes. The mold, along with the mixture, was then compressed at 700 MPa and placed in an isothermal oven at 65 °C for sintering for 1.5 h. Pure PLGA beads were simultaneously prepared as a control.

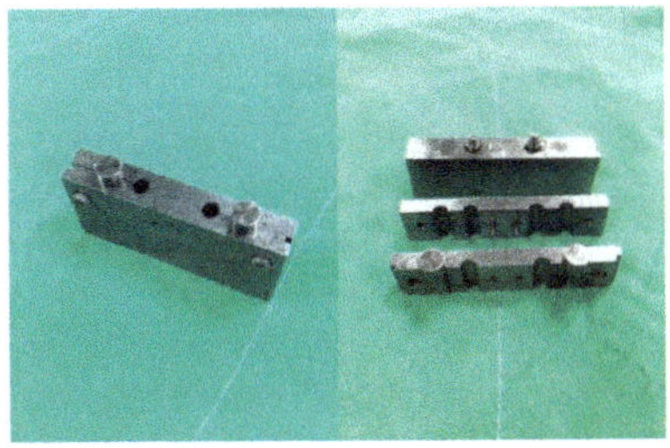

**Figure 1.** Mold used for compression molding of the antifungal-agent-containing beads.

**Table 1.** Composition of the fluconazole-loaded Poly(D,L-lactide-*co*-glycolide) beads.

| PLGA/Drug Ratio | Size: Diameter × Height (mm × mm) | Weight of PLGA (mg) | Weight of Drug (mg) |
|---|---|---|---|
| 6:1 | 3 × 5.65 | 54 | 9 |
| | 5 × 5.55 | 150 | 25 |
| | 8 × 6.18 | 384 | 64 |
| 4:1 | 3 × 5.02 | 50.4 | 12.6 |
| | 5 × 6.27 | 140 | 35 |
| | 8 × 6.25 | 358.4 | 89.6 |

### 2.2. *In Vitro Analysis of Amphotericin B and Fluconazole Release*

In vitro release of amphotericin B and fluconazole from the biodegradable beads was assessed via an elution method. PLGA/amphotericin B and PLGA/fluconazole beads were placed in test tubes ($N = 3$) containing 1 mL buffered solution at 37 °C. The tubes were deposed in an isothermal oven for 24 h, and the eluent was harvested and substituted with fresh solution (1 mL). The process was carried out in duplicate for 40 d. Amphotericin B and fluconazole levels in the harvested eluents were quantified via HPLC, carried out using a Hitachi L-2200 System (Hitachi Medical Systems, Tokyo, Japan). All experiments were performed in triplicate ($N = 3$).

### 2.3. *FTIR Spectrometry*

The thermal stability of fluconazole was determined via FTIR spectrometry to evaluate whether the chemistry and orientation of material structures varied with temperature. FTIR analysis was conducted using a Nicolet iS5 spectrometer (Thermo Fisher Scientific, Waltham, MA, USA) at a resolution of 4 $cm^{-1}$ (32 scans). The drugs were compressed as KBr discs, and spectra were recorded over 400–4000 $cm^{-1}$. Fluconazole was considered stable if no obvious structural changes occurred due to temperature. The FTIR spectra of pure PLGA beads were compared with those of fluconazole-loaded PLGA beads.

### 2.4. *Determination of the Water Contact Angle*

The water contact angles of the beads were measured on a contact angle measurement device (First Ten Angstroms, Portsmouth, VA, USA) ($N = 3$).

### 2.5. *Cell Culture*

Cytotoxicity of fluconazole-loaded PLGA beads was examined via a Cell Counting Kit-8 (CCK-8) assay (Sigma-Aldrich, St. Louis, MO, USA) for cell viability, in accordance with the manufacturer's instructions. Eluents harvested at 1, 2, 3, 7, and 14 d were placed in 96-well culture plates. Human fibroblasts obtained from foreskins of patients (1–3 years of age) undergoing surgery were seeded ($1 \times 10^4$ cells/well) in Dulbecco's Modified Eagle's Medium (DMEM) at 37 °C and 5% $CO_2$/95% for 48 h. Cell viability was monitored via the CCK-8 assay and quantified using an ELISA reader.

### 2.6. *In Vivo Animal Study*

All animal experimental procedures were approved by the Institutional Animal Care and Use Committee of Chang Gung University (CGU107-275, Approved 19 March 2019), and all experimental animals were provided care in accordance with the regulations of the Ministry of Health and Welfare of Taiwan under the supervision of a licensed veterinarian.

Four adult New Zealand white rabbits (Animal Health Research Institute, Panchiao, Taiwan) weighing approximately 3.5 ± 0.3 kg were enrolled in the experiment. The rabbits were housed in individual cages in a temperature- and light-controlled room, with ad libitum access to standard rabbit chow and sterilized drinking water. All animals were administered general anesthesia via inhalation of isoflurane through a vaporizer (Matrx, Pompano Beach, FL, USA) in a plastic box (40 cm × 20 cm × 28 cm). Anesthesia was maintained during the entire surgical procedure via mask inhalation of isoflurane.

After rabbits were sedated, the right femoral sites were depilated, washed with soft soap, and treated aseptically with 70% ethanol directly before the surgical procedure. The other site of the animals was covered with a sterile blanket. Under aseptic conditions, the middle/third region of the right femur was dissected via an anterolateral approach. A bone defect ($5.0 \times 10.0$ mm$^2$) was induced at the right femoral middle site, and a polymethylmethacrylate (PMMA) spacer was initially inserted. The wound was closed with 3-0 Vicryl sutures (Johnson & Johnson, New Brunswick, NJ, USA). After 2 weeks, the PMMA spacer was surgically excised and fluconazole-impregnated PLGA cylindrical beads ($5.0 \times 6.0$ mm$^2$) were placed into the right femoral bone cavity (Figure 2). The wound was closed in a layer-by-layer manner. In vivo drug concentrations were determined by sampling specimens (surgical site fluid) with aspirates obtained on days 1, 2, 3, 7, and 14. After 2 weeks, owing to difficulty aspirating fluid from the surgical site, we harvested tissue around the bead area on days 21, 28, 35, 42, and 49. Cylindrical specimens of tissue surrounding the beads were centrifuged, and the plasma was sampled and stored at −80 °C until analysis. In vivo fluconazole concentrations in the tissue samples were determined via HPLC. All samples were diluted with phosphate-buffered saline and assessed in accordance with the assay standard curve. A calibration curve was generated for each set of measurements (correlation coefficient > 0.99). Blood samples were obtained from the marginal ear vein using a syringe to determine serum aspartate transaminase (AST) levels and antibiotic concentrations after implanting the fluconazole-impregnated beads. AST levels were determined using the IDEXX Catalyst DX system (Westbrook, ME, USA), while fluconazole concentrations were determined via HPLC. Data thus obtained were used to assess the liver function of rabbits during experiments.

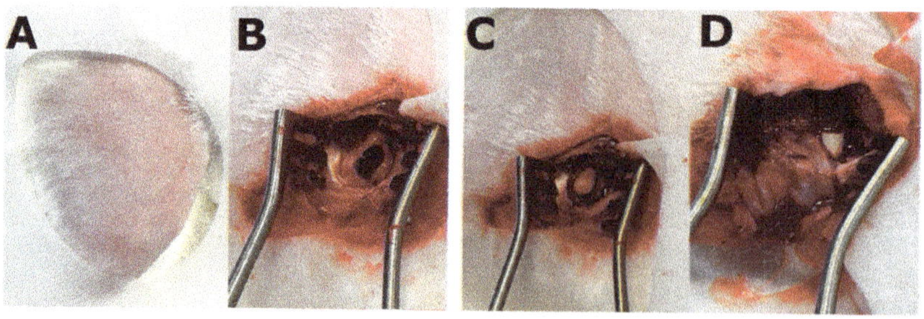


**Figure 2.** Images of the surgical procedure. (**A**) The right femoral site was depilated and sterilized. (**B**) a bone cavity was formed ($5.0 \times 10.0$ mm$^2$); (**C**) a polymethylmethacrylate spacer was placed into the bone cavity; (**D**) after 2 weeks, fluconazole-impregnated Poly(D,L-lactide-*co*-glycolide) cylindrical beads ($5.0 \times 6.0$ mm$^2$) were placed into the right femoral bone cavity.

## 3. Results

### *3.1. Characterization of Fabricated Poly(D,L-lactide-co-glycolide) PLGA/Amphotericin B and PLGA/Fluconazole Beads*

Amphotericin B- and fluconazole-containing beads of 3, 5, and 8 mm were prepared through compression molding (Figure 3). To confirm that the molding temperature did not deactivate the drugs, a thermal stability test for fluconazole was carried out using FTIR spectroscopy. As shown in Figure 4, amphotericin B and fluconazole remained intact at 70 °C, indicating that a temperature of 65 °C is optimal for compression molding.

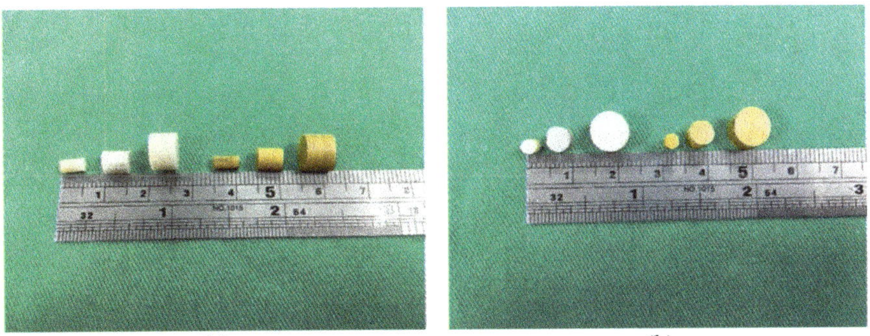

(a) (b)

**Figure 3.** Photographs of the fabricated drug-loaded beads. The white cylindrical beads are fluconazole-impregnated Poly(D,L-lactide-*co*-glycolide) (PLGA) beads. The yellow cylindrical beads are amphotericin B-impregnated PLGA beads.

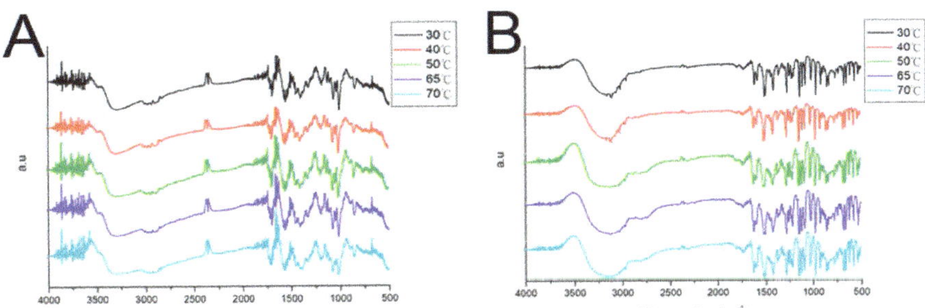


**Figure 4.** Effect of molding temperature on (**A**) amphotericin B and (**B**) fluconazole stability determined by Fourier-transform infrared spectroscopy.

Water contact angles determined herein are shown in Figure 5. While the pure PLGA beads exhibited hydrophobic properties (water contact angle of 97.38°), antifungal drug-containing beads were hydrophilic (all angles were less than 70°). In addition, the water contact angle of PLGA/fluconazole beads decreased with an increase in the drug content of beads, primarily owing to the hydrophilic nature of fluconazole.

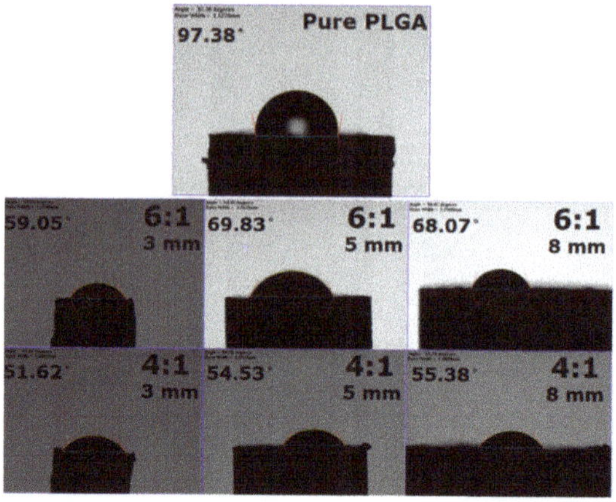


**Figure 5.** Water contact angles for pure Poly(D,L-lactide-*co*-glycolide) and drug-loaded beads of different polymer:drug ratios and sizes.

To confirm successful incorporation of amphotericin B and fluconazole in the beads, FTIR spectra of drug-loaded beads were compared with those of pure PLGA beads (Figure 6). The new absorption peak at 1600 $cm^{-1}$ might be attributable to the C=N bonds of amphotericin B and fluconazole [18,19]. Enhanced absorption at 1670–1780 $cm^{-1}$ corresponded to C=O bonds, primarily owing to supplementation with antifungal drugs, and an absorbance peak of approximately 1270 $cm^{-1}$ may have resulted from C–F bond enhancement in loaded drugs. The FTIR spectra indicate that the antifungal drugs were successfully incorporated into PLGA beads.

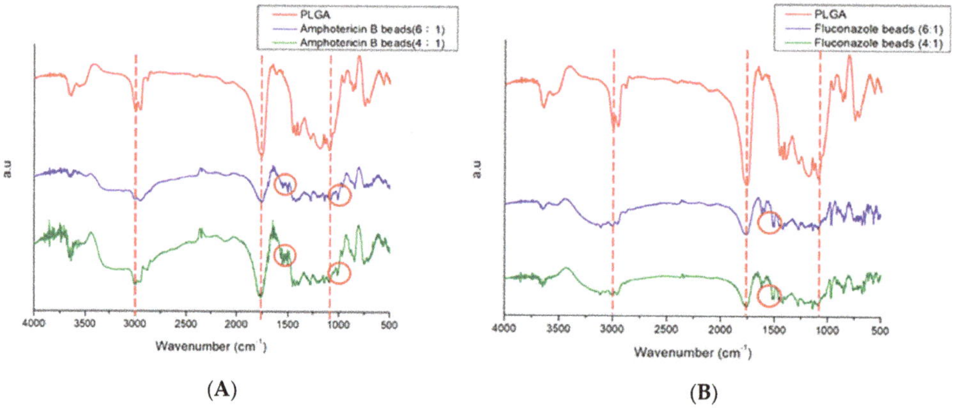


**Figure 6.** Fourier-transform infrared spectra of (**A**) pure Poly(D,L-lactide-*co*-glycolide) (PLGA) and amphotericin B/PLGA beads (**B**) pure Poly(D,L-lactide-*co*-glycolide) (PLGA) and fluconazole/PLGA beads.

### *3.2. In Vitro Release Dynamics of Poly(D,L-lactide-co-glycolide) PLGA/Amphotericin B and PLGA/Fluconazole Beads*

Figure 7 illustrates the release dynamics of amphotericin B and fluconazole from antifungal-drug-loaded beads with different polymer:drug ratios (6:1 and 4:1). Drug release was slightly less than 1% of the total release of amphotericin B. Triphasic liberation curves were generated, displaying blast release on the first day, followed by gradual elution on days 2–23, and accelerated drug release on days 23 and 10 in beads measuring 3, 5, and 8 mm, respectively. Beads with a higher drug loading ratio (i.e., polymer:drug = 4:1) generally exhibited greater antifungal drug concentrations in the eluent. Figure 8 shows the release dynamics of PLGA/fluconazole beads of different sizes. Larger beads exhibited an earlier accelerated release of fluconazole than smaller beads. All antifungal-drug-embedded beads exhibited sustained release of fluconazole for more than 30 d.

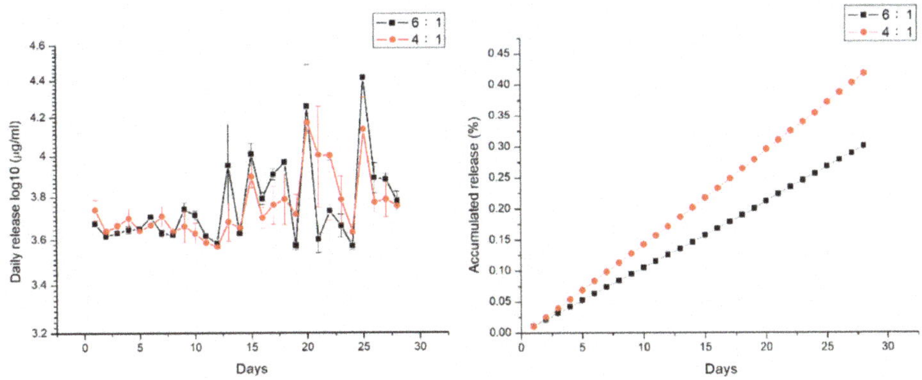


**Figure 7.** In vitro release curves of amphotericin B from the drug-loaded beads.

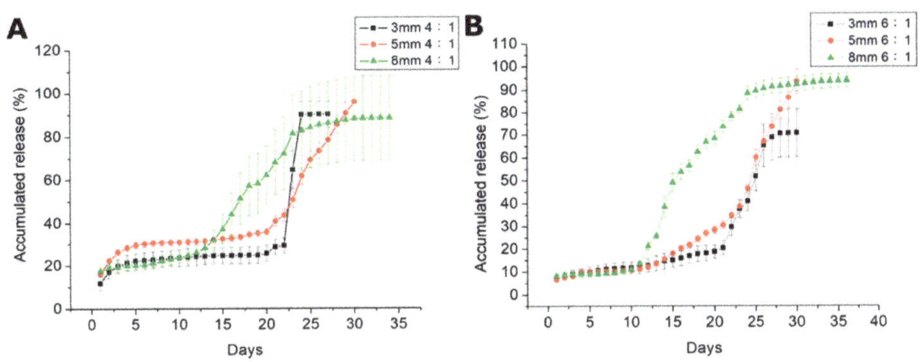


**Figure 8.** In vitro release curves of fluconazole from drug-loaded beads with various polymer:drug ratios. (**A**). The accumulated release of beads of different sizes with 4:1 polymer:drug ratios. (**B**). The accumulated release of beads of different sizes with 6:1 polymer:drug ratios.

Cytotoxicity analysis was performed using CCK-8 assays to measure cell viability. The eluent from 5 mm beads with a 4:1 polymer:drug ratio was analyzed. Figure 9 shows that cell viability was reduced on day 1, probably owing to burst release of fluconazole, thereby potentially affecting cell proliferation. Thereafter, PLGA/fluconazole beads showed no signs of cytotoxicity.

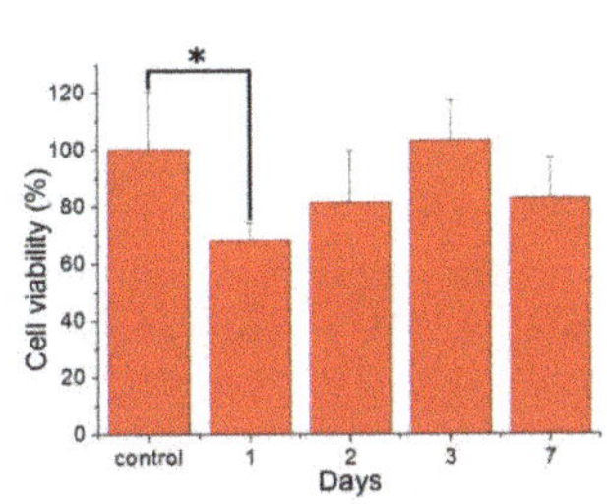


**Figure 9.** Cell viability of the fluconazole-incorporated beads (* $p < 0.05$).

### 3.3. *In Vivo Drug Release*

Owing to the poor release dynamics of amphotericin B, only PLGA/fluconazole beads were selected for in vivo analysis of drug release. Figure 10 shows drug concentrations as measured in bone cavity tissue and blood. Fluconazole-embedded beads displayed long-term fluconazole release (beyond the MTC) for more than 49 d in vivo. Furthermore, blood drug concentration was significantly lower than that in bone tissue.

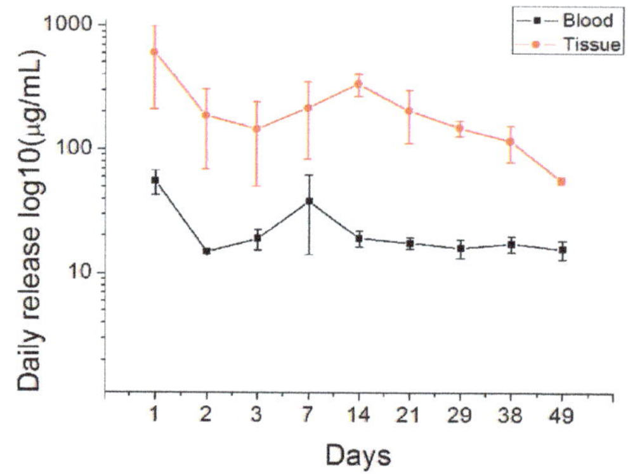


**Figure 10.** In vivo release curves of fluconazole from the drug-loaded beads.

Blood AST levels were within the physiological range (Figure 11) of 33–99 (U/L) [20].

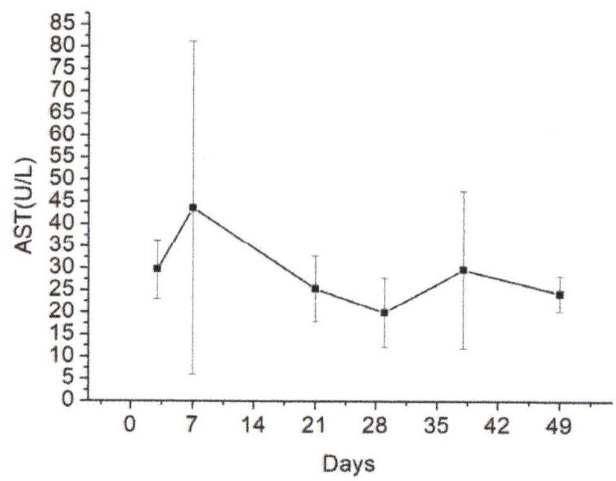


**Figure 11.** Estimation of blood aspartate transaminase (AST) levels.

## 4. Discussion

Treatment of fungal osteomyelitis is more complicated than that of chronic osteomyelitis. Local antibiotic release is important in treating chronic osteomyelitis; however, the use of antifungal-loaded bone cement beads to treat fungal osteomyelitis remains controversial, mainly owing to the inconsistent release dynamics of antifungal drugs from the PMMA beads. While some studies have reported the successful eradication of fungal osteomyelitis or periprosthetic joint infections [21], others have yielded conflicting results. During production of PMMA beads containing amphotericin B or fluconazole, covalent crosslinkage may result in poor drug release from the bone cement [13,14]. The potential risk of irritation of host tissues and the non-biodegradable nature of PMMA potentially limit its applications in treating osteomyelitis, and therefore, surgical excision is often performed as an alternative. This study reports the successful production of antifungal-agent-loaded PLGA beads with amphotericin B or fluconazole using the compression-molding method. In vitro analysis revealed that the elution rate of amphotericin-B-loaded PLGA beads with different ratios did not exceed 1% for 28 d (Figure 7). Moreover, amphotericin-B-loaded PLGA beads displayed poor sustained drug release. However, the fluconazole-loaded PLGA beads with different ratios displayed an elution rate above 90% for 28 d. Based on these results, we selected PLGA/fluconazole beads for in vivo analysis.

Few studies have evaluated the efficacy of antifungal-impregnated PLGA carriers [22–24], although the release profile of amphotericin B with an organic solvent was shown to be promising. In the present study, PLGA/amphotericin B beads showed poor release dynamics [25]. The temperature during compression molding was increased up to 65 °C, which is also the temperature used for polymerization of amphotericin-impregnated PMMA beads, before sintering for 1.5 h. The polymerization process may have facilitated covalent linkages in amphotericin-impregnated PMMA beads; however, the actual mechanism resulting in the poor release profile of amphotericin-impregnated PLGA beads remains unclear. Nonetheless, poor release was not observed with fluconazole-impregnated PLGA beads.

Biodegradable PLGA/fluconazole beads released antifungal drugs at high concentrations for over 49 d, thus controlling the bone infection. The present work is the first study to develop biodegradable antifungal beads using a compression-molding technique without the use of organic solvents and to evaluate the sustained release of high and local fluconazole concentrations in vivo. Owing to the absence of organic solvents during bead preparation, these fluconazole/PLGA beads are potentially applicable for clinical use for the treatment of fungal infections in bone tissue. The present results show that most drugs were absorbed by the surrounding tissues, while the systemic drug concentration remained low (Figure 10).

Fluconazole is a first-generation triazole antifungal agent used to treat fungal infections through both oral and parental routes. Fluconazole is metabolized in the kidneys, unlike other azoles metabolized in the liver. Adverse effects have been commonly reported with use of long-term fluconazole therapy [26], and although none of these have been severe or life-threatening [19], liver damage has been observed in some cases [27,28]. In a previous study, elevation of liver transaminases was more common than liver damage, and the incidence of treatment termination owing to elevated liver enzymes was 0.7% [28]. In the present study, hepatoxicity potentially resulting from fluconazole in the blood during the treatment period was assessed upon local administration at high doses. Blood samples were collected from the marginal ear vein to quantify blood AST levels after each biopsy. Louie et al. reported an optimal fluconazole therapeutic concentration of not more than 10 mg/kg [29]. In the present study, fluconazole was used at 10 mg/kg (a cylindrical bead containing 35 mg fluconazole/3.5 kg body weight), and blood AST levels were within the physiological range in all the New Zealand white rabbits used in the study. These results indicate that fluconazole-impregnated PLGA beads are safe (Figure 11).

PLGA is one of the most suitable biodegradable polymeric materials for synthesizing drug delivery devices and for tissue engineering [30–32]. The material is biocompatible and biodegradable, exhibits a wide range of erosion times, and is mechanically tunable. Most importantly, PLGA is an FDA-approved polymer used extensively for controlled release of small-molecule drugs, proteins, and other macromolecules. Therefore, we selected PLGA as the carrier material for antifungal fluconazole-containing beads.

Drug release from a biodegradable carrier generally occurs in three stages: primary blast, diffusion-dominated elution, and degradation-dominated release [15]. After compression molding, most drugs are dispersed into the volume of the PLGA/fluconazole beads; however, certain drug formulations on the particle surface may lead to an initial drug release burst, followed by controlled drug release by diffusion and other factors. Relatively constant slow elution of the antifungal agents was thus observed. Finally, PLGA/fluconazole beads swell owing to water uptake during elution, thus damaging the polymer matrix and forming openings for antifungal release. The rate of fluconazole release thus accelerated accordingly.

Sustained local release of high levels of antifungals contributes to infection control. Louie et al. reported that the minimum inhibitory concentration median ($MIC_{50}$) in a designated fluconazole mid-resistant infection was 64–128 μg/mL. In a fluconazole-susceptible strain of *Candida*, the median MIC was 0.5 μg/mL [29], and the concentration of released fluconazole was greater than 128 μg/mL. These results demonstrate that biodegradable antifungal-embedded beads can release high concentrations of fluconazole (well beyond the $MIC_{50}$) for more than 49 d (Figure 10). Furthermore, FTIR analysis suggests that bead-embedded drug formulations remain stable during the molding process.

Although the current study has generated promising preliminary data, some limitations should be noted. First, we used a non-infected animal model, and therefore it is unclear whether the antifungal beads might perform differently in infected tissue. Further evaluation of the antifungal agent-embedded PLGA copolymer beads in an animal model of fungal infection is necessary to address this limitation. Second, despite the experimental data showing that 80% of the drug was released at the end of the study, it would have been faster given larger amounts of liquid. The sink condition analysis is needed to ensure the drug was released freely. Third, although no obvious sign of inflammation was observed in the in vivo test, the influence of controlled release of drugs and carriers on the local irritation should be further examined. Finally, the relevance of the present findings to patients with bone infections remains unclear and warrants further investigation. We intend to further explore these topics in future studies.

## 5. Conclusions

This study is the first to perform in vitro and in vivo analyses of drug release dynamics in the bone cavity from compression-molded antifungal-incorporated PLGA beads prepared without organic solvents. Biodegradable PLGA/fluconazole beads released high concentrations of antifungals for over 49 d at the target site, while the antifungal agent blood concentration remained low. In summary, the compression-molded PLGA/fluconazole beads that we describe here have potential applications for treating bone infections.

**Author Contributions:** Y.-H.H., S.W.-N.U., and S.-J.L.; methodology, Y.-H.H., H.-Y.C., J.-C.C., and Y.-H.Y.; validation, Y.-H.H., S.W.-N.U., and S.-J.L.; formal analysis, Y.-H.H., H.-Y.C., J.-C.C.; investigation, Y.-C.C.; resources, Y.-H.H. and S.-J.L.; data curation, Y.-H.H., H.-Y.C., J.-C.C., and S.-J.L.; writing—original draft preparation, Y.-H.H.; writing—review and editing, Y.-H.H. and S.-J.L.; visualization, Y.-H.Y. and Y.-C.C.; supervision, S.W.-N.U. and S.-J.L.; project administration, Y.-H.H., H.-Y.C., and J.-C.C.; funding acquisition, S.W.-N.U. and S.-J.L.

**Funding:** This work was supported by the Ministry of Science and Technology, Taiwan (Contract No. 107-2221-E-182-017) and the Chang Gung Memorial Hospital (Contract No. CMRPD2H0032).

**Conflicts of Interest:** The authors declare no conflict of interest.

## References

1. Gamaletsou, M.N.; Walsh, T.J.; Sipsas, N.V. Epidemiology of fungal osteomyelitis. *Curr. Fungal Infect. Rep.* **2014**, *8*, 262–270. [CrossRef]
2. Gamaletsou, M.N.; Rammaert, B.; Bueno, M.A.; Moriyama, B.; Sipsas, N.V.; Kontoyiannis, D.P.; Roilides, E.; Zeller, V.; Prinapori, R.; Taj-Aldeen, S.J.; et al. Aspergillus osteomyelitis: Epidemiology, clinical manifestations, management, and outcome. *J. Infect.* **2014**, *68*, 478–493. [CrossRef] [PubMed]
3. Gamaletsou, M.N.; Kontoyiannis, D.P.; Sipsas, N.V.; Moriyama, B.; Alexander, E.; Roilides, E.; Brause, B.; Walsh, T.J. *Candida* osteomyelitis: Analysis of 207 pediatric and adult cases (1970–2011). *Clin. Infect. Dis.* **2012**, *55*, 1338–1351. [CrossRef] [PubMed]
4. Desai, J.V.; Mitchell, A.P.; Andes, D.R. Fungal biofilms, drug resistance, and recurrent infection. *Cold Spring Harb. Perspect. Med.* **2014**, *4*, a019729. [CrossRef] [PubMed]
5. Rammaert, B.; Gamaletsou, M.N.; Zeller, V.; Elie, C.; Prinapori, R.; Taj-Aldeen, S.J.; Roilides, E.; Kontoyiannis, D.P.; Brause, B.; Sipsas, N.V.; et al. Dimorphic fungal osteoarticular infections. *Eur. J. Clin. Microbiol. Infect. Dis.* **2014**, *33*, 2131–2140. [CrossRef]
6. Bariteau, J.T.; Waryasz, G.R.; McDonnell, M.; Fischer, S.A.; Hayda, R.A.; Born, C.T. Fungal osteomyelitis and septic arthritis. *J. Am. Acad. Orthop. Surg.* **2014**, *22*, 390–401. [CrossRef]

7. Mader, J.T.; Calhoun, J.; Cobos, J. In vitro evaluation of antibiotic diffusion from antibiotic-impregnated biodegradable beads and polymethylmethacrylate beads. *Antimicrob. Agents Chemother.* **1997**, *41*, 415–418. [CrossRef]
8. Henry, M.W.; Miller, A.O.; Walsh, T.J.; Brause, B.D. Fungal musculoskeletal infections. *Infect. Dis. Clin. N. Am.* **2017**, *31*, 353–368. [CrossRef]
9. Cornely, O.A.; Lasso, M.; Betts, R.; Klimko, N.; Vazquez, J.; Dobb, G.; Velez, J.; Williams-Diaz, A.; Lipka, J.; Taylor, A.; et al. Caspofungin for the treatment of less common forms of invasive candidiasis. *J. Antimicrob. Chemother.* **2007**, *60*, 363–369. [CrossRef]
10. Marcos-Zambrano, L.J.; Escribano, P.; Bouza, E.; Guinea, J. Comparison of the antifungal activity of micafungin and amphotericin B against *Candida tropicalis* biofilms. *J. Antimicrob. Chemother.* **2016**, *71*, 2498–2501. [CrossRef]
11. Pappas, P.G.; Kauffman, C.A.; Andes, D.R.; Clancy, C.J.; Marr, K.A.; Ostrosky-Zeichner, L.; Reboli, A.C.; Schuster, M.G.; Vazquez, J.A.; Walsh, T.J.; et al. Clinical practice guideline for the management of candidiasis: 2016 update by the Infectious Diseases Society of America. *Clin. Infect. Dis.* **2016**, *62*, e1–e50. [CrossRef] [PubMed]
12. Kweon, C.; McLaren, A.C.; Leon, C.; McLemore, R. Amphotericin B delivery from bone cement increases with porosity but strength decreases. *Clin. Orthop. Relat. Res.* **2011**, *469*, 3002–3007. [CrossRef] [PubMed]
13. Sealy, P.I.; Nguyen, C.; Tucci, M.; Benghuzzi, H.; Cleary, J.D. Delivery of antifungal agents using bioactive and nonbioactive bone cements. *Ann. Pharmacother.* **2009**, *43*, 1606–1615. [CrossRef] [PubMed]
14. Goss, B.; Lutton, C.; Weinrauch, P.; Jabur, M.; Gillett, G.; Crawford, R. Elution and mechanical properties of antifungal bone cement. *J. Arthroplast.* **2007**, *22*, 902–908. [CrossRef] [PubMed]
15. Liu, S.J.; Ueng, S.W.N.; Chan, E.C.; Lin, S.S.; Tsai, C.H.; Wei, F.C.; Shih, C.H. In vitro elution of vancomycin from biodegradable beads. *J. Biomed. Mater. Res. (Appl. Biomater.)* **1999**, *48*, 613–620. [CrossRef]
16. Amass, W.; Amass, A.; Tighe, B. A review of biodegradable polymers: Uses, current developments in the synthesis and characterization of biodegradable polyesters, blends of biodegradable polymers and recent advances in biodegradation studies. *Polym. Int.* **1998**, *47*, 89–144. [CrossRef]
17. Doppalapudi, S.; Jain, A.; Khan, W.; Domb, A.J. Biodegradable polymers—An overview. *Biodegrad. Polym.* **2014**, *25*, 427–435. [CrossRef]
18. Gagoś, M.; Arczewska, M. FTIR spectroscopic study of molecular organization of the antibiotic amphotericin B in aqueous solution and in DPPC lipid monolayers containing the sterols cholesterol and ergosterol. *Eur. Biophys. J.* **2012**, *41*, 663–673. [CrossRef]
19. Atay, O.; Selck, F. Quantitative determination of fluconazole by infrared spectrophotometry. *Anal. Lett.* **1996**, *29*, 2163–2176. [CrossRef]
20. Burnett, N.; Mathura, K.; Metivier, K.S.; Holder, R.B.; Brown, G.; Campbell, M. An investigation into haematological and serum chemistry parameters of rabbits in Trinidad. *World Rabbit Sci.* **2006**, *14*, 175–187. [CrossRef]
21. Wu, M.H.; Hsu, K.Y. Candidal arthritis in revision knee arthroplasty successfully treated with sequential parenteral-oral fluconazole and amphotericin B-loaded cement spacer. *Knee Surg. Sports Traumatol. Arthrosc.* **2011**, *19*, 273–276. [CrossRef] [PubMed]
22. Khan, A.A.; Alanazi, A.M.; Jabeen, M.; Khan, S.; Malik, A. Additive potential of combination therapy against cryptococcosis employing a novel amphotericin B and fluconazole loaded dual delivery system. *Eur. J. Pharm. Sci.* **2018**, *119*, 171–178. [CrossRef] [PubMed]
23. Souza, A.O.; Nascimento, A.L.; de Vasconcelos, N.M.; Jerônimo, M.S.; Siqueira, I.M.; R-Santos, L.; Cintra, D.O.; Fuscaldi, L.L.; Pires Júnior, O.R.; Titze-de-Almeida, R.; et al. Activity and in vivo tracking of Amphotericin B loaded PLGA nanoparticles. *Eur. J. Med. Chem.* **2015**, *95*, 267–276. [CrossRef] [PubMed]
24. Amaral, A.C.; Bocca, A.L.; Ribeiro, A.M.; Nunes, J.; Peixoto, D.L.; Simioni, A.R.; Primo, F.L.; Lacava, Z.G.; Bentes, R.; Titze-de-Almeida, R.; et al. Amphotericin B in poly(lactic-co-glycolic acid) (PLGA) and dimercaptosuccinic acid (DMSA) nanoparticles against paracoccidioidomycosis. *J. Antimicrob. Chemother.* **2009**, *63*, 526–533. [CrossRef] [PubMed]
25. Reckling, F.W.; Dillon, W.L. The bone-cement interface temperature during total joint replacement. *J. Bone Joint Surg. Am.* **1977**, *59*, 80–82. [CrossRef] [PubMed]
26. Davis, M.R.; Nguyen, M.V.; Donnelley, M.A.; Thompson, G.R., III. Tolerability of long-term fluconazole therapy. *J. Antimicrob. Chemother.* **2019**, *74*, 768–771. [CrossRef] [PubMed]

27. Tverdek, F.P.; Kofteridis, D.; Kontoyiannis, D.P. Antifungal agents and liver toxicity: A complex interaction. *Expert Rev. Anti Infect. Ther.* **2016**, *14*, 765–776. [CrossRef]
28. Wang, J.L.; Chang, C.H.; Young-Xu, Y.; Chan, K.A. Systematic review and meta-analysis of the tolerability and hepatotoxicity of antifungals in empirical and definitive therapy for invasive fungal infection. *Antimicrob. Agents Chemother.* **2010**, *54*, 2409–2419. [CrossRef]
29. Louie, A.; Banerjee, P.; Drusano, G.L.; Shayegani, M.; Miller, M.H. Interaction between fluconazole and amphotericin B in mice with systemic infection due to fluconazole-susceptible or -resistant strains of *Candida albicans*. *Antimicrob. Agents Chemother.* **1999**, *43*, 2841–2847. [CrossRef]
30. Stewart, R.L.; Cox, J.T.; Volgas, D.; Stannard, J.; Duffy, L.; Waites, K.B.; Chu, T.M. The use of a biodegradable, loadbearing scaffold as a carrier for antibiotics in an infected open fracture model. *J. Orthop. Trauma.* **2010**, *24*, 587–591. [CrossRef]
31. Williams, D.F. Biodegradation of surgical polymers. *J. Mater. Sci.* **1982**, *17*, 1233–1246. [CrossRef]
32. Kumbar, S.G.; Nukavarapu, S.P.; James, R.; Nair, L.S.; Laurencin, C.T. Electrospun poly(lactic acid-co-glycolic acid) scaffolds for skin tissue engineering. *Biomaterials* **2008**, *29*, 4100–4107. [CrossRef] [PubMed]

*pharmaceutics*

MDPI

*Article*

# Development and Characterization of a Semi-Solid Dosage Form of Meglumine Antimoniate for Topical Treatment of Cutaneous Leishmaniasis

**Diana Berenguer [1], Lilian Sosa [2], Magdalena Alcover [1], Marcella Sessa [2], Lyda Halbaut [2], Carme Guillén [1], Roser Fisa [1], Ana Cristina Calpena-Campmany [2] and Cristina Riera [1,*]**

1 Department of Biology, Health and Environment, Laboratory of Parasitology, Faculty of Pharmacy and Food Sciences, University of Barcelona, 08028 Barcelona, Spain; berenguer.diana@gmail.com (D.B.); mmagdalenaalcoveramengual@ub.edu (M.A.); carmeguillen@ub.edu (C.G.); rfisa@ub.edu (R.F.)

2 Department of Pharmaceutical Technology and Physicochemistry, Faculty of Pharmacy and Food Sciences, University of Barcelona, 08028 Barcelona, Spain; liliansosa2012@gmail.com (L.S.); marcellass93@gmail.com (M.S.); halbaut@ub.edu (L.H.); anacalpena@ub.edu (A.C.C.-C.)

* Correspondence: mcriera@ub.edu; Tel.: +34-93-4024500

Received: 10 October 2019; Accepted: 13 November 2019; Published: 15 November 2019

**Abstract:** Cutaneous leishmaniasis (CL) is treated with painful intralesional injections of meglumine antimoniate (MA). With the aim of developing an alternative topical treatment for CL, a gel-based formulation with 30% MA was prepared and its physicochemical properties, stability and rheological behavior were studied. The following were assessed: drug release on propylene hydrophilic membranes ex vivo human skin permeation, tolerance in healthy volunteers, cytotoxicity in three cell lines and anti-leishmanial activity against *Leishmania infantum* promastigotes and amastigotes. The MA gel formulation was found to have suitable pH, and good spreadability and stability. Low quantities of pentavalent antimony ($Sb^V$) were observed in release and permeation tests, whereas retention was high in both non-damaged and damaged skin (71,043.69 ± 10,641.57 and 10,728 ± 2254.61 μg/g/cm$^2$ of $Sb^V$, respectively). The formulation did not have a toxic effect on the cell lines, and presented lower $Sb^V$ $IC_{50}$ values against amastigotes (15.76 ± 4.81 μg/mL) in comparison with the MA solution. The high amount of drug retained in the skin and the $Sb^V$ $IC_{50}$ values obtained suggest that this semi-solid dosage form has potential as an alternative treatment of CL.

**Keywords:** *Leishmania infantum*; cutaneous leishmaniasis; meglumine antimoniate; Sepigel 305®; topical treatment

## 1. Introduction

Leishmaniasis is caused by protozoan parasites, including more than 20 *Leishmania* species, and is transmitted through the bites of infected female phlebotomine sandflies. It is a neglected tropical disease that is prevalent in low-income countries [1] and is endemic in 98 countries with tropical, temperate and mild temperate climates, with an estimated 12 million people infected by the disease. There are three types of leishmaniasis: cutaneous (CL), mucocutaneous (MCL) and visceral (VL). The most common form of the disease worldwide is CL, with 0.6–1 million new cases estimated to occur annually [2]. In the Mediterranean area, *Leishmania infantum* is the main species responsible for CL and VL [3]. Depending on the *Leishmania* species involved, manifestations of CL range from self-resolving ulcerative lesions, which may take up to two years to heal, to painful open wounds that leave permanent scars or even disseminate and affect other areas of the body [4].

Treatments proposed for CL depend on the infecting species, patient immunity, as well as the number, sizes and location of lesions, but none of them are universally effective [5]. CL cases presenting small and localized lesions are usually treated with pentavalent antimony salts, sodium stibogluconate or meglumine antimoniate (MA). These first-line drugs are administered by intralesional infiltrations requiring multiple injections that cause acute pain to the extent that patients are often unable to complete the treatment [6]. As it is a common mistake when talking about MA treatments for leishmaniasis to confuse the therapeutic dose that refers to MA with the amount of pentavalent antimony ($Sb^V$), henceforth we will refer all the concentrations to the amount of $Sb^V$ since it is the main active ingredient.

Consequently, efforts have been made to develop alternative drugs and forms of administration.

According to the World Health Organization (WHO), topical and local therapies with less systemic toxicity (thermotherapy, cryotherapy, paromomycin ointment, local infiltration with antimonials) are good treatment options for patients with a low number of small lesions [7].

Topical treatments have many advantages: they are easy to administer, do not require hospitalization or other infrastructure, favor patient commitment, are low cost and have fewer toxic side effects [8–10].

A widely used treatment is the application of 15% paromomycin ointments in association with 12% methylbenzethonium chloride twice a day for 20 days, although results are variable, and patients experience significant irritation, inflammation, pain and pigmentation [11,12]. Cure rates of around 80% have been obtained in some clinical trials [13,14], whereas others report lower efficiency [15]. Liposomes and a nanogel formulation containing paromomycin have also been developed [6,16,17].

Alternatively, topical conjugates with amphotericin B were reported to penetrate the skin in a murine model [18]. However, the cure rates in a phase II study with topical 3% amphotericin B cream for uncomplicated CL were low and its clinical development as a therapeutic option was discontinued [19].

Clinical trials have also been performed with a topical formulation of miltefosine, but no efficacy against CL has been proven [10] despite experimental trials in mice reporting a reduction in lesion size and full recovery, albeit with relapse [20].

Regarding antimonials, there are few data available on semi-solid dosage forms for topical administration. The development and characterization of liposomes loading sodium stibogluconate and MA has been described [21–23].

Sepigel 305®, a product used in gels and gel-creams, consists of a gelling agent (polyacrylamide), a non-ionic emulsifier (polyoxyethylene 7 lauryl ether) and a fatty oil (isoparaffin). Simple gel formulations containing Sepigel 305® have medium to high viscosity [24], a degree of cooling power, evanescent properties and optimal dermocosmetic qualities. They also allow the incorporation of both hydrophilic and lipophilic substances, and previous studies on different drugs have reported that formulations containing this polymer have good stability [24–26].

In this context, the use of hydrogels could be a promising strategy to deliver MA for the management and healing of CL. The aim of the present study was therefore to design and develop a gel-based formulation containing MA for CL treatment. Physico-chemical parameters were monitored to determine stability over time. Human skin permeation and retention ex vivo and tolerance in vivo were tested to optimize a new dosage system of the drug with fewer side effects. Cytotoxic effects and leishmanicidal activity in vitro were assayed in promastigotes and intracellular amastigotes of *L. infantum*.

## 2. Materials and Methods

### 2.1. Materials

Meglumine antimoniate was obtained from Acros Organics (Thermo Fisher Scientific, Waltham, MA, USA). Gentamicin was acquired from Sigma-Aldrich (Darmstadt, Germany), and Sepigel 305® from Acofarma (Barcelona, Spain). Distilled water utilized in the experiments was obtained from a Mili-Q® Plus System (Millipore Co., Burlington, MA, USA).

## 2.2. Preparation of the Gel

Briefly, a gel-based formulation containing 30% of MA (with 31.6% $Sb^V$ *w/w*) was prepared. MA was dissolved in water with 0.5% of gentamicin. Sepigel 305® was added under continuous stirring to obtain the gel formulation at a concentration of 4.5%.

## 2.3. Physicochemical Characterization of the MA Gel

### 2.3.1. Morphological Analysis

To analyze its morphology, the MA gel was dried over a period of 7 days using a vacuum desiccator. Once totally dried, 0.1 g was coated with carbon as a conductive agent. The internal structure of the gel was examined by scanning electron microscopy (SEM) using a JEOL J-7100FE (Peabody, MA, USA).

### 2.3.2. Swelling and Degradation Tests

A gravimetric method was used to obtain the swelling ratio (SR) and to test degradation, which was represented as the percentage of weight loss (WL). Dried and fresh MA gel were used to carry out the swelling and degradation tests, respectively. In both experiments the MA gel was incubated in PBS (pH = 5.5) at 32 °C for 20 min. Samples ($n$ = 6) were removed and weighed after blotting the surface water at predetermined time intervals of 3 min. The SR was calculated using the following equation and expressed by kinetic modeling:

$$\text{SR} = \frac{Ws - Wd}{Wd} \tag{1}$$

where $Ws$ is the weight of the swollen MA gel at 3 min intervals for 20 min and $Wd$ is the weight of dried gel.

WL was calculated following the equation and expressed by kinetic modeling:

$$\text{WL}\ (\%) = \frac{Wi - Wd}{Wi} \times 100 \tag{2}$$

where $Wi$ is the initial weight of the MA gel and $Wd$ is the gel weight each 3 min.

### 2.3.3. Water Loss Due to Drying

Water loss was evaluated by weighing 1 g of fresh MA gel, which was placed in a vacuum desiccator for 7 days until the weight was constant.

### 2.3.4. Porosity Study

The porosity percentage (P) was calculated by a solvent displacement method, which consisted of immersing the previously dried MA gel in absolute ethanol for 2 min and then weighing it after the excess ethanol on the surface was blotted. The porosity percentage was calculated using the following equation:

$$\text{P}\ (\%) = \frac{\text{W2} - \text{W1}}{\rho \times V} \times 100 \tag{3}$$

where $W1$ represents the weight of the dried MA gel, $W2$ stands for the weight of the MA gel after being immersed in ethanol, $\rho$ is the density of absolute ethanol and $V$ is the volume of the gel.

## 2.4. Stability Studies

The MA gel was stored at room temperature (RT), 4 °C and 37 °C. The pH values of the formulation were measured with universal test paper (Filter-Lab®) by direct spread of the samples ($n$ = 3). Appearance was visually inspected, and pH was measured after 30 days and after 6 months.

$Sb^V$ was quantified by ICP-OES (Inductively coupled plasma-optical emission spectrometry, Perkin Elmer Elan 6000, Waltham, MA, USA) after previous treatment of the samples with $HNO_3$ and

$H_2O_2$ in a microwave digester at 220 °C for 72 h. The drug content was evaluated at the beginning and end of the 6-month experiment.

### 2.5. Rheological Studies

The rheological measurements were performed in duplicate 24 h after gel preparation using a rotational rheometer (Thermo Scientific HaakeRheostress 1, Thermo Fischer Scientific, Karlsruhe, Germany) equipped with a cone plate set-up (0.105 mm gap between cone and plate) with a fixed lower plate and a mobile upper cone Haake C60/2° Ti (60 mm diameter, 2° angle). The rheometer was connected to a temperature control ThermoHaake Phoenix II + Haake C25P (Thermo Fischer Scientific, Waltham, MA, USA) and a computer provided with the HaakeRheowin® Job Manager v. 4.0 (Thermo Fischer Scientific, Waltham, MA, USA) to carry out the tests and HaakeRheowin® Data Manager v.4.0 (Thermo Fischer Scientific) to carry out the analyses of the obtained data. Viscosity curves ($\eta = f(\dot{\gamma})$) and flow curves ($\tau = f(\dot{\gamma})$) were recorded at 25 °C. The shear rate ramp program included: a ramp-up period from 0 to 50 $s^{-1}$ for 3 min; constant shear rate period of 50 $s^{-1}$ for 1 min; and a ramp-down period from 50 to 0 $s^{-1}$ for 3 min. The data from the flow curves ($\tau = f(\dot{\gamma})$) were fitted to different mathematical models equations: Newton, Bingham, Ostwald-de-Waele, Herschel-Bulkley, Casson and Cross [27]. Best fit of mathematical models was based on the correlation coefficient value (r). The viscosity mean value (Pa·s) was determined from the constant share section at 50 $s^{-1}$.

### 2.6. Spreadability Test

The spreadability of MA gel and MA solution was determined in triplicate as follows: 0.5 g of formulation and 150 µL of MA solution were placed within 1 cm diameter circles previously marked on a glass plate, and also on a plastic plate. A series of weights (5, 10, 20, 50 and 100 g) were successively added and allowed to rest for 2 min each at RT. The diameters ($cm^2$) of the circle spreads were measured and recorded as comparative values.

### 2.7. In Vitro Release Studies

Vertical Franz diffusion cells (FDC 400; Crown Glass, Somerville, NJ, USA) were used to assay the MA release with hydrophilic polypropylene membranes (GH Polypro, Life Sciences). The receptor medium was water at 32 ± 0.5 °C under stirring at 600 rpm in sink conditions. The experiment was performed with a diffusion area of 0.64 $cm^2$. Small quantities of MA gels (0.3 g) or MA solution (300 µL) were added to the donor compartment. At the end of the study at 55 h, the amount of $Sb^V$ in the receptor compartment was analyzed by ICP-OES (Perkin Elmer Elan 6000). Results were reported as the mean ± SD of five replicates.

### 2.8. Ex Vivo Permeation Studies

Ex vivo permeation tests were performed as described in Section 2.7 using damaged and non-damaged human skin samples with a thickness of 400 µm from a single donor. The skin was damaged using the tape stripping technique, which involves repeated application of adhesive tapes to the skin surface to remove stratum corneum layers. For the assay, adhesive tapes were applied to the skin 7 times to damage the skin but not remove it completely. Then 0.3 g of MA gel was placed in the donor compartment. After 27 h the quantity of $Sb^V$ was determined by ICP-OES in five replicates.

To determine the amount of drug retained in the epidermal tegument, the skin membranes were detached from the Franz cells, cleaned with gauze soaked in 0.05% sodium dodecylsulphate solution and washed in distilled water. The area of permeation was then cut and weighed. Finally, the $Sb^V$ retained was extracted in an ultrasound bath with water for 30 min. The resulting solutions were measured by ICP-OES for quantification.

The Plastic Surgery Department of Barcelona-SCIAS Hospital (Barcelona, Spain) provided the human skin from the abdominal region of a healthy woman, and the Bioethics Committee of the same hospital approved the experimental protocol (reference number: BEC/001/16; date: 15 January 2016).

Written informed consent forms were provided. Human skin was dermatomed (GA630 dermatome, Aesculap, Tuttlingen, Germany) into pieces with a thickness of 400 μm [28]. The integrity of skin samples was assessed in triplicate by measuring the trans-epidermal water loss (TEWL) values using a DermaLab® module (Cortex Technology, Hadsund, Denmark).

### 2.9. *In Vivo Tolerance Study*

Ten female volunteers with healthy skin between 25 and 35 years old participated in the study. The study was approved by the Ethics Committee of the University of Barcelona (reference number: IRB00003099; date: 20 March 2018) following the guidelines of the Declaration of Helsinki [29] and all volunteers signed written informed consent forms. Skin-care cosmetics were not permitted on the test areas for two days prior to the study. Volunteers stayed in the test room for at least 30 min prior to the measurements. Measurements were performed before applying the gel (to establish the baseline readings) and 15 min, 1 h and 2 h after the application of 0.5 g on the flexor side of the left forearm.

Trans epidermal water loss (TEWL), referring to the total amount of water vapor lost through the skin was measured by a Tewameter® TM 300 (Courage-Khazaka electronic GmbH, Cologne, Germany). The stratum corneum hydration (SCH) was determined using a Corneometer® CM 825 (Courage-Khazaka electronic GmbH). Measurements were performed according to international guidelines [30].

### 2.10. *Parasite Strains and Cultures*

The *Leishmania infantum* strain MHOM/ES/2016/CATB101 isolated from an individual with CL from Mallorca (Spain) was used. Promastigotes were cultured at 26 °C in Schneider insect medium, pH 7.0, with 20% heat-inactivated fetal calf serum, 25 μg/mL gentamicin solution (Sigma G-1397, St. Louis, MO, USA), and 1% penicillin (100 U/mL)-streptomycin (100 mg/mL) solution (Sigma P4333).

### 2.11. *In Vitro Cytotoxicity Assay*

In order to study the cytotoxicity of the MA gel and the gel excipients, $5 \times 10^4$ cells/mL of the cell lines RAW 264.7, J774A.1 and HaCaT were cultured in 96-well plates (Costar 3595). Serial dilutions of the gel in RPMI-1640 medium with 10% heat-inactivated fetal calf serum and 1% penicillin (100 U/mL)-streptomycin (100 mg/mL) were added. After 24 h of incubation at 37 °C in a 5% $CO_2$ atmosphere, WST-1 reagent (Roche Diagnostics GmbH, Mannheim, Germany) was added to each well and the plate was incubated in the same conditions for 4 h. The absorbance of the samples was measured at 450 nm using a spectrophotometer (Multiskan EX, Thermo Electron Corporation, Shanghai, China). The concentration inhibiting 50% of cell viability ($CC_{50}$) was determined by linear regression analysis and experiments were performed in triplicate [31].

### 2.12. *In Vitro Anti-Leishmanial Activity against Promastigotes*

The activity of MA gel, gel alone and MA solution was studied on promastigotes in microtiter plates (Costar 3595). Serial dilutions of the gel in Schneider medium were seeded, and then $10^6$ promastigotes/mL in their logarithmic growth phase were added to each well and incubated at 26 °C for 48 h. Growth was measured through acid phosphatase and the optical density was determined at 405 nm with a spectrophotometer (Multiskan EX, Thermo Electron Corporation). The concentration inhibiting 50% of parasite growth ($IC_{50}$) was determined by linear regression analysis of the minimum squares of parasitic growth versus the logarithm of the drug concentration with 95% confidence interval. Experiments were performed in triplicate [32].

### 2.13. *In Vitro Anti-Leishmanial Activity against Intracellular Amastigotes*

J774A.1 cells at a concentration of $5 \times 10^4$ cells/mL were cultured in an eight LabTek chamber slide system (Nunc®). After 24 h, late stationary phase promastigotes from a 5-day culture in RPMI-1640 complete medium with 10% heat-inactivated fetal calf serum and 1% penicillin (100 U/mL)-streptomycin (100 mg/mL) were added at a concentration of 5 × $10^5$ cells/mL and incubated for 24 h at 37 °C in 5% $CO_2$ atmosphere. Free promastigotes were removed by washing. RPMI-1640 complete medium with serial dilutions of the gel was added to each well and incubated at 37 °C in 5% $CO_2$ atmosphere for 48 h. Infected cells were washed and slides were stained with Giemsa. Drug activity was evaluated by counting the number of infected cells by examining 300 macrophages in triplicate [33].

### 2.14. *Statistical Analysis*

Experimental data obtained were analyzed by one-way parametric analysis of variance (ANOVA), followed by a multiple comparison Tukey test. A $p < 0.05$ indicated the differences were statistically significant. Prism® V. 5 (GraphPad Software Inc., San Diego, CA, USA) was used for calculations.

## 3. Results

### 3.1. *Physicochemical Characterization of the MA Gel*

As shown in Figure 1A,B, SEM revealed a laminar disposition in the gel formulation. The layers were regularly organized and a porous structure was evident.

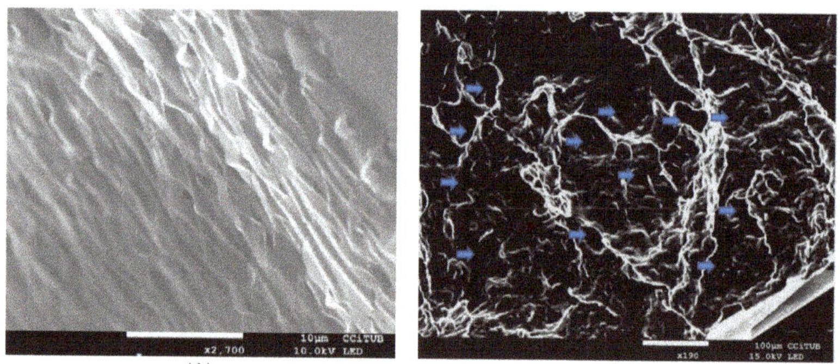


**Figure 1.** SEM images of the meglumine antimoniate (MA) gel, (**A**) laminar disposition (2700×) and (**B**) porous structure (190×).

The swelling process of MA gel followed a hyperbolic model, which was represented by the kinetic constants k = 0.64 $min^{-1}$ ($r^2$ = 0.9986) (Figure 2A). The degradation of MA gel was completed in 15 min and followed a one phase exponential model with a kinetic constant k = 0.01 $min^{-1}$ ($r^2$ = 0.9995) (Figure 2B), and the P percentage of MA gel was ~89.37 ± 0.15%.

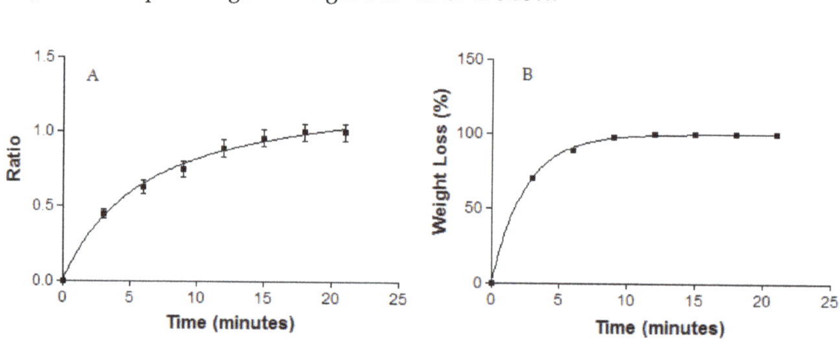


**Figure 2.** (**A**) Swelling ratio and (**B**) percentage of weight loss in the degradation of MA gel.

After 7 days of drying the gel, the sample showed 22 ± 2.13% loss of water content.

### 3.2. Drug Quantification and Stability Studies

Determination of the drug content revealed 72.7% entrapment efficiency of $Sb^V$ in the gel formulation. The concentration of $Sb^V$ in the reference solution and in the gel over time is summarized in Table 1. The $Sb^V$ content of the gel samples at the tested temperatures did not present statistical differences ($p > 0.05$) after 6 months (Figure 3).

**Table 1.** Concentration of $Sb^V$ and SD in the reference solution and MA gel samples at the beginning of the experiment and 6 months later at different storage temperatures ($t_0$ = initial time; $t_{180}$ = after 6 months).

| Formulations | $Sb^V$ (µg/mL) ± SD | | | |
|---|---|---|---|---|
| | $t_0$ | $t_{180}$ RT | $t_{180}$ 37 °C | $t_{180}$ 4 °C |
| MA solution | 79,659.92 ± 1023.45 | | | |
| MA gel | 57,936.62 ± 894.92 | 57,436.62 ± 854.92 | 58,575.93 ± 654.07 | 59,255.83 ± 524.49 |

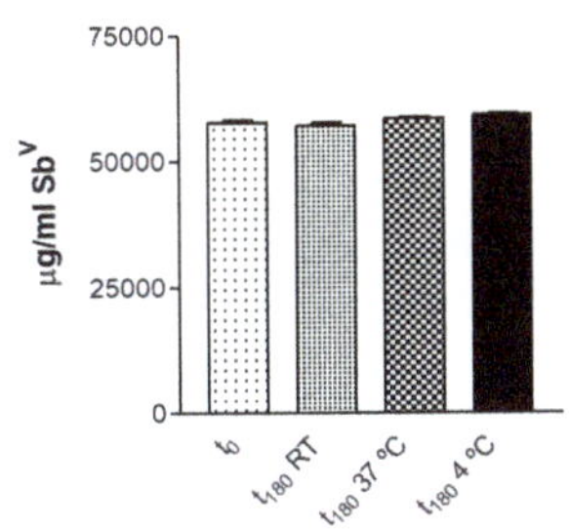


**Figure 3.** Concentration of $Sb^V$ in the gel at different storage temperatures at the beginning of the experiment and 6 months later.

The MA gel had a pH value of 5–6 at the beginning and end of the experiment.

When freshly prepared, the gel formulation was a semi-transparent white with homogeneous appearance without precipitates. After 30 days at RT and 4 °C, the formulation had the same appearance as at the beginning of the experiment, whereas at 37 °C it turned yellow. After 6 months at 4 °C it looked the same as at the beginning of the experiment, while at RT it was a semi-transparent light yellow and at 37 °C a semi-transparent dark yellow.

### 3.3. Rheological Studies

The MA gel formulation exhibited pseudoplastic behavior (Cross model; $r^2 = 0.9997$) (Figure 4A). Viscosity at 50 $s^{-1}$ was 677.8 ± 1.76 mPa·s, and it was non-thixotropic. The MA solution showed Newtonian behavior (Newton model; $r^2 = 0.9967$) (Figure 4B) and viscosity at 50 $s^{-1}$ was 2.67 ± 0.06 mPa·s.

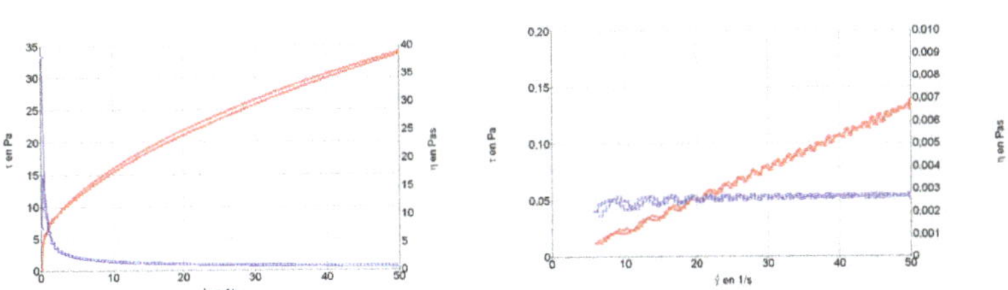

(A) (B)

**Figure 4.** Rheological behavior of (**A**) MA gel and (**B**) MA solution.

### 3.4. Spreadability Test

The equation that best fitted the spreadability behavior of the MA gel and solution followed a first-order kinetic model [$S = S_{max} \times (1 - \exp(-k \times W))$], where S is the extension surface ($cm^2$), $S_{max}$ is the maximum extension surface ($cm^2$), k is the kinetic constant ($g^{-1}$) and W is the weight added (g). The resulting graphics and equations are shown in Figure 5. Spreadability of the MA gel increased as more weight was applied, until a maximum value was reached.

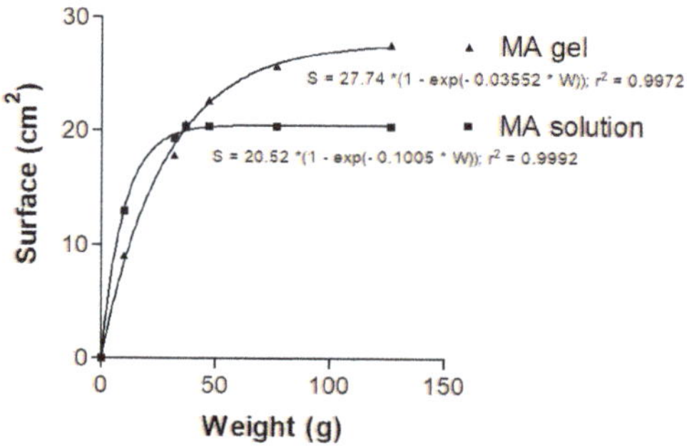


**Figure 5.** Spreadability graphic and first-order kinetic model equations for the MA gel and MA solution.

### 3.5. In Vitro Release Studies

After 55 h the amount of $Sb^V$ released in the receptor compartment through the hydrophilic polypropylene membranes was 13,646.53 ± 960.63 μg/cm² of $Sb^V$ for MA solution and 9081.13 ± 1446.20 μg/cm² of $Sb^V$ for the gel, corresponding to a release of 32.3 ± 3.56% and 21.5 ± 3.43%, respectively, from the total amount seeded. The differences ($p < 0.01$) between the $Sb^V$ released from the solution and the gel are statistically significant (Figure 6).

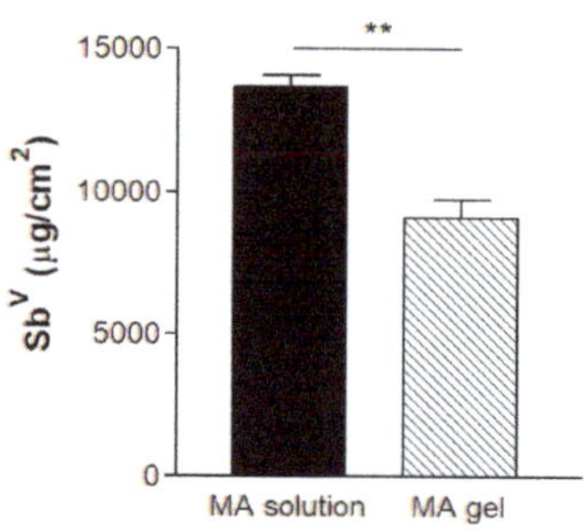


**Figure 6.** Flux of $Sb^V$ released after 55 h through hydrophilic polypropylene membranes (** $p < 0.01$).

### 3.6. Ex Vivo Permeation Studies

Table 2 summarizes the amount of $Sb^V$ permeated (μg/cm²) through the skin and the amount retained (μg/g/cm²). In both permeation and retention assays, higher $Sb^V$ values were obtained for non-damaged than damaged skin when testing the reference solution and the MA gel. The percentage of $Sb^V$ in the reference solution and MA gel that permeated through the healthy skin was 7.03 ± 1.23% and 2.89 ± 0.83%, and through the damaged skin 0.29 ± 0.51% and 1.84 ± 0.65%, respectively.

**Table 2.** $Sb^V$ permeation across the skin barrier and $Sb^V$ retained after 27 h in damaged and non-damaged skin.

| Assay | MA Solution | | MA Gel | |
|---|---|---|---|---|
| | **Non-Damaged Skin** | **Damaged Skin** | **Non-Damaged Skin** | **Damaged Skin** |
| Permeation (μg/cm$^2$) ± SD | 2966.50 ± 562.37 | 121.00 ± 59.45 | 1217.53 ± 279.41 | 774.81 ± 179.63 |
| Retention in skin (μg/g/cm$^2$) ± SD | 51,672.84 ± 8964.28 | 2057.47 ± 381.67 | 71,043.69 ± 10,641.57 | 10,728.23 ± 2254.61 |

### 3.7. *In Vivo Tolerance Study*

The results for the TEWL and SCH are shown in Figure 7. The MA solution decreased the TEWL with statistically significant differences for all of the experiment times (Figure 7A). The MA gel increased the TEWL after 15 min, and basal values were almost recovered after 1 h, with statistically significant differences between 15 min and 60 min, and between 15 min and 120 min (Figure 7C). As the graph shows, significant differences in SCH values were found when comparing baseline values with all the experiment times for both formulations, revealing the dehydration of the stratum corneum and its slow gradual recovery (Figure 7B,D).

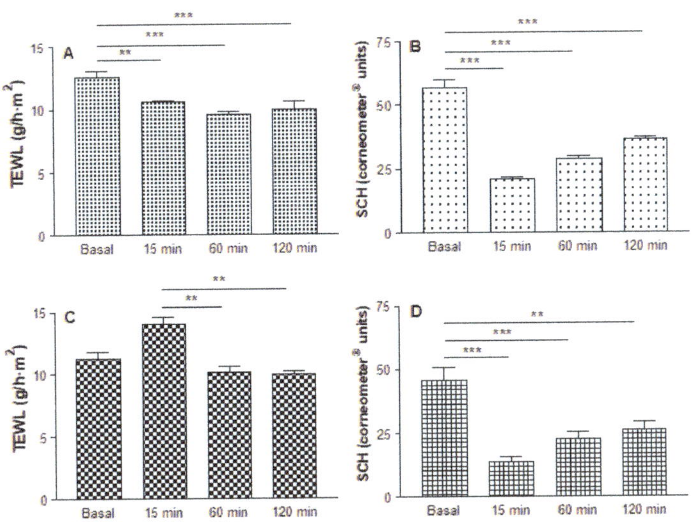


**Figure 7.** Measurements of biomechanical parameters of skin for 2 h after application of the gel and solution. (**A**) Change in trans-epidermal water loss (TEWL) and (**B**) Stratum corneum hydration (SCH) values after application of MA solution. (**C**) Change in TEWL and (**D**) SCH values after application of MA gel (** $p < 0.01$; *** $p < 0.001$).

### 3.8. *In Vitro Cytotoxicity Assay*

Figures 8 and 9 show the cytotoxic effects on the cell lines assayed. No cytotoxicity was observed in the keratinocyte cell line HaCaT at the ranges of dilution tested. The gel formulation with and without MA showed cytotoxicity in the RAW264.7 and J774A.1 cell lines only at the highest concentrations of the gelling excipients and $Sb^V$ assayed. The $CC_{50}$ of the MA gel and the gel alone are summarized in Tables 3 and 4.

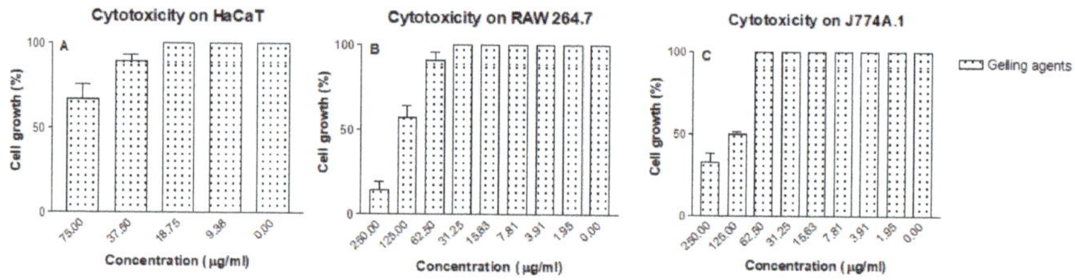


**Figure 8.** Cytotoxicity of the gelling excipients in cell lines (**A**) HaCaT, (**B**) RAW 264.7 and (**C**) J774A.1.

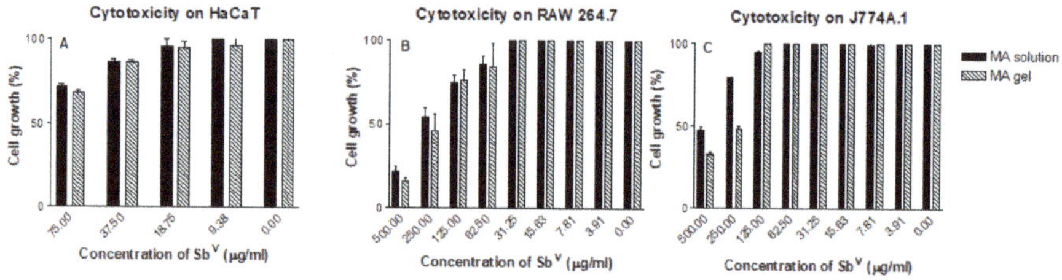


**Figure 9.** Cytotoxicity of MA solution and MA gel in cell lines (**A**) HaCaT, (**B**) RAW 264.7 and (**C**) J774A.1.

**Table 3.** In vitro activity against promastigotes and amastigotes, and cytotoxicity of the reference MA solution and the MA gel and selectivity index (SI).

| Formulations (μg/mL Sb$^V$) | $IC_{50}$ (μg/mL Sb$^V$ ± SD) | | SI | | $CC_{50}$ (μg/mL Sb$^V$ ± SD) | |
|---|---|---|---|---|---|---|
| | **Promastigotes** | **Amastigotes** | **$SI_{RAW}$** | **$SI_{J774}$** | **RAW 264.7** | **J774A.1** |
| MA solution | >750 (* na) | 57.35 ± 2.76 | 6.66 | 6.37 | 381.76 ± 94.74 | 365.14 ± 165.26 |
| MA gel | 633.15 ± 43.26 (* na) | 15.76 ± 4.81 | 14.19 | 14.03 | 223.66 ± 46.82 | 221.05 ± 65.41 |

* na: not active.

**Table 4.** In vitro activity against promastigotes and amastigotes, and cytotoxicity of the gel alone and selectivity index (SI).

| Excipients | $IC_{50}$ (μg/mL ± SD) | | SI | | $CC_{50}$ (μg/mL ± SD) | |
|---|---|---|---|---|---|---|
| | **Promastigotes** | **Amastigotes** | **$SI_{RAW}$** | **$SI_{J774}$** | **RAW 264.7** | **J774A.1** |
| Gelling agents (375–0.37 μg/mL) | >375 | 20.39 ± 5.43 | 6.57 | 6.13 | 134.00 ± 42.57 | 125.00 ± 20.80 |

### *3.9. Anti-Leishmanial In Vitro Activity against Promastigotes*

The effects of the MA solution and MA gel are summarized in Table 3, and Table 4 shows the effect of the gel without the drug. Neither the MA gel nor the gel alone were able to inhibit the growth of *L. infantum* promastigotes, with Sb$^V$ $IC_{50}$ values of 633.15 ± 43.26 μg/mL (0.70% of gel) for the MA gel and $IC_{50}$ values greater than 375 μg/mL (0.71% of gel) for the gel alone.

### *3.10. Anti-Leishmanial In Vitro Activity against Intracellular Amastigotes*

The activity of the MA solution and MA gel is summarized in Table 3. In the amastigote assay the MA gel presented lower Sb$^V$ $IC_{50}$ values (15.76 ± 4.81 μg/mL) in comparison with the MA solution (57.35 ± 2.76 μg/mL). As shown in Table 4, the gelling excipients showed some activity against the amastigotes of *L. infantum*. Figure 10 shows the images of control infected macrophages and infected macrophages treated with MA gel.

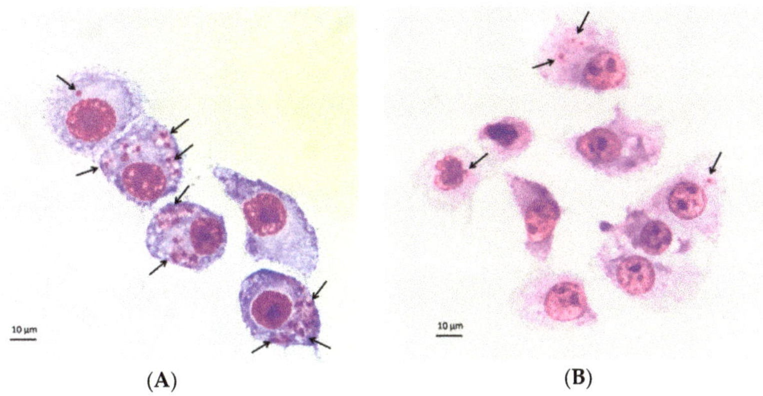


**Figure 10.** Effect of the MA gel on the infection of J774A.1 macrophages with *L. infantum*. (**A**) Infected and untreated control cultures. (**B**) Cultures treated with 15.63 µg/mL of $Sb^V$ of the MA gel. Amastigotes are indicated by arrows.

## 4. Discussion

The development of an effective topical treatment for CL is an interesting but difficult challenge. One of the main problems in topical administration is that *Leishmania* amastigotes are localized within dermal macrophages, in the papillary dermis beneath the epidermis [34]. Thus, after the drug is released from the vehicle, it must cross the stratum corneum in the skin barrier and reach the macrophages to deliver the pharmacological action.

In this study, we developed a Sepigel 305®-based formulation for the dermal delivery of MA with qualities that could enhance the activity of MA against CL with minimal side effects.

The water content in the MA gel formulation represented almost a quarter of its composition. Degradation was constant over time, indicating it did not depend on the remaining polymer concentration, and the process was completed in 15 min. The very high porosity percentage of the MA gel was corroborated by SEM images, which also revealed a regular laminar disposition in the superficial structure of the gel layers. The porous structure, similar to that described by Dong et al. (2015) [35], allows the drug to be released when applied on membranes or skin.

The pH plays an important role in the stability of gels as well as in skin tolerability, and should be between 5 and 6 for skin formulations [26]. The tested formulation maintained a pH of 5–6 throughout the experiment, indicating biocompatibility for skin applications, with minimal risk of irritation or bacterial or fungal infection [36]. Anchisi et al. (2001) studied the stability of a wide range of formulations containing the same gel base over 6 months, and observed minimal changes in the pH [24].

At the beginning of the experiment the quantity of $Sb^V$ found in the gel was 57,936.62 µg/mL, 27.3% less than in the reference solution consisting of MA in water. This reduction may be explained by incomplete digestion during the analytical procedure, considering the high quantity of organic material, which may mask the actual amount of $Sb^V$. The stability test indicated that the amount of $Sb^V$ in the gels remained constant for 6 months at all the tested storage temperatures, although the organoleptic characteristics varied. Samples stored at 4 °C retained the same visual properties throughout the 6 months (a homogeneous semi-transparent white appearance), whereas samples stored at RT turned pale yellow in the third month, and at 37 °C they turned yellow after the first month. This change in color may be due to meglumine oxidation with increasing temperature. Gel homogeneity was observed in all samples, without any biphasic formation.

When comparing different topical formulations, Bilia et al. (2006) found the best long-term stability with polyacrylamide, C13-14 isoparaffin and laureth-7 [25]. Similar results were reported by

Anchisi et al. (2001), who observed that samples containing Sepigel 305® did not polymerize at high temperatures because of the ability of polyacrylamide to trap free water; organoleptic features of the formulations were also stable throughout the six-month study period [24].

The rheological properties of topical semi-solid dosage forms are important because they influence skin spreadability, adhesion and retention in the application site [21]. The rheological analysis indicated that the MA gel is a pseudoplastic fluid, like most gels reported in the literature, which is ideal for topical administration [37]. As such, its main characteristic is decreased viscosity when the deformation speed is increased, resulting in a good spreading capacity. The pseudoplastic behavior is determined by the cross linking of the polymer. The MA solution showed Newtonian behavior due to its predominant water composition and very low viscosity, which makes it difficult to avoid spilling or to maintain skin adhesion after application.

Spreadability, which is considered a quality parameter for skin formulations, was lost in the MA solution at 37 g, but continued to increase in the more viscous MA gel until 127 g, indicating its ease of application (see rheograms in Figure 4A,B).

The profile of drug release from the delivery vehicle can be useful for predicting in vivo behavior of the product. At 55 h, drug release from the formulation was incomplete: only 32.3 ± 3.56% of $Sb^V$ from the MA solution and 21.5 ± 3.43% from the MA gel had been released. This was probably due to the large amount of sample located in the donor compartment, the slow release of the drug from the gel, and the chemical binding of the meglumine.

In addition, the permeation study performed with damaged and healthy human skin showed very low amounts of $Sb^V$ in the receptor compartment at 27 h for both the MA solution and MA gel. The percentage of permeation was higher through healthy skin than damaged skin with both vehicles. In the retention study, the drug in the skin was determined in terms of µg of $Sb^V$ per g and $cm^2$ of skin. Higher $Sb^V$ retention values were obtained for the MA gel when the skin was undamaged, probably due to better functioning and ability to absorb more drug. The same behavior was observed for the MA solution, although a lower amount of $Sb^V$ was detected. Similar results were obtained by Dar et al. (2018), who tested nano-deformable liposomes with sodium stibogluconate and the same liposomes in a Carbopol gel on rat skin, and observed a low percentage of $Sb^V$ permeated into the receptor compartment and a high percentage was retained in the skin [21]. Other studies with liposomes encapsulating MA, MA combined with oleic acid and MA plus stearylamine also showed low permeation of $Sb^V$ across the skin barrier and high retention in mouse skin [22,23]. Bilia et al. (2006) found the highest amounts of sesquiterpenes retained in stripped skin when using a mixture of polyacrylamide, C13-14 isoparaffin and laureth-7 and other excipients, since the level of lipophilicity enhanced the permeation compared with other topical delivery systems [25].

Most substances penetrate the skin through the transepidermic route, mainly across the spaces between cells [38]. The level of penetration depends essentially on two coefficients: that of the vehicle-skin partition and the diffusion. Small molecules are able to cross the skin more easily [39], while highly lipophilic molecules can be retained in the stratum corneum [40]. The drug of our formulation may cross the skin through the intercellular route, as MA is a small molecule (MW: 365.98 g/mol), thereby allowing proper diffusion despite its hydrophilicity. In the permeation experiments, we found lower $Sb^V$ values in the damaged skin than when it was not. This indicates that stratum corneum plays an important role as a drug reservoir that allows a slow release, as we could detect $Sb^V$ in the receptor compartment. Regarding the new formulation, the low permeation of $Sb^V$ to the receptor compartment through the skin barrier, and retention of sufficient quantities of $Sb^V$ in the damaged skin after 27 h suggests the drug could remain in the skin layers where leishmania parasites are located.

In relation to biomechanical properties, the MA solution maintained TEWL values below 12 $g/h{\cdot}m^2$, which is considered normal, and the integrity of the stratum corneum was not affected. The slight decreasing pattern observed could be due to an occlusive effect of the solution. The solution also caused a loss of water in the skin but with a tendency to recover after 1 h. When applying the MA gel, an increase in water loss from the stratum corneum was observed 15 min after the application, pointing to an alteration in the constitution of the skin. However, within one hour the basal values were recovered, which indicated the alterations were reversible. When using gels containing polymers with a high capacity to capture water, the skin hydration is strongly affected and water is lost from the stratum corneum. In the current study, this effect was marked at 15 min, followed by a slow gradual recuperation of hydration over the next 2 h. The application of the MA gel did not induce any visual skin irritation and was well tolerated by all the individuals participating in the study, none of whom reported pain or any inconveniences from its administration.

At the highest concentrations tested, the gel formulation with or without MA showed cytotoxicity effects on the RAW 264.7 and J774A.1 cell lines. Cytotoxicity in the macrophage cell lines was similar for all the compounds tested, indicating a similar resistance to the excipients and the MA in the formulation.

The in vitro anti-leishmanial activity on intracellular amastigotes showed the MA gel inhibited amastigote survival to a greater extent than the MA solution. Analysis of the selectivity index (SI) showed higher SI values for the MA gel formulation than for the MA solution, which indicated that the gel was less toxic and with higher activity against the parasite.

Interestingly, the amount of $Sb^V$ retained in damaged and undamaged skin was greater when the gel formulation was applied compared to the solution, which makes it more advantageous for therapeutic application. These skin retention values, as well as the $IC_{50}$ results against amastigotes (Table 3) and considering the density of the hydrated skin (0.964 g/mL) [41], indicate that the new MA gel formulation could provide high local amounts of $Sb^V$, even in damaged skin, and may exert an efficient action against CL. Future experiments in an animal model need to be performed to confirm our preliminary results and determine the efficacy in vivo.

## 5. Conclusions

We have developed an easy to prepare topical MA gel-based formulation, an administration route that avoids injections and has higher patient acceptance and tolerability. It was shown to be stable for at least 6 months and has optimal properties for use in CL treatment: a biocompatible pH, easy spreadability and no skin irritant effects.

Experiments showed low permeation of the $Sb^V$ through the skin, which would represent a low concentration of the drug in blood, and high retention in the skin layers. Thus, the gel formulation almost completely avoids the toxic side effects of $Sb^V$ and could supply a therapeutic level of MA at the local site for an extended period.

Low cytotoxicity in the keratinocyte cell line and the two macrophages cell lines was observed. The study of anti-leishmanial activity in intracellular amastigotes treated with the MA gel showed a reduction of the $IC_{50}$ in comparison to the reference solution. These preliminary results indicate the MA formulation could be a promising alternative for CL treatment, although further studies are required to establish a proper dosage schedule in vivo.

**Author Contributions:** D.B. carried out and analyzed the experiments and wrote the original draft of the manuscript. L.S. conceptualized the idea, carried out the swelling, degradation and porosity tests, and edited the manuscript. M.A. and C.G. collaborated in the preparation of the parasitology experiments. M.S. carried out and analyzed the experiments. L.H. directed the rheological tests. R.F. reviewed the manuscript. A.C.C.-C. conceptualized the idea of the investigation and validated and supervised the biopharmaceutical results. C.R. conceptualized the idea, validated and supervised the parasitology experiments, reviewed and edited the manuscript.

**Funding:** This research received no external funding.

**Acknowledgments:** The authors would like to thank the University of Barcelona for the financial support to cover the cost of open access publication. We appreciate the help of Silvia Tebar for the preparation of culture mediums. We would like to thank the group of Josefa Badia, and specially Rodrigo Vera for helping us with the assay on keratinocyte cell line. We value the contribution of Joan Garcias for his expertise in dermatology. Additionally, thanks to Lucy Brzoska for the revision of the use of the English language.

**Conflicts of Interest:** The authors declare no conflict of interest.

## References


1. Weng, H.B.; Chen, H.X.; Wang, M.W. Innovation in neglected tropical disease drug discovery and development. *Infect. Dis. Poverty* **2018**, *7*, 67–76. [CrossRef] [PubMed]
2. Alvar, J.; Vélez, I.D.; Bern, C.; Herrero, M.; Desjeux, P.; Cano, J.; Janin, J.; den Boer, M.; WHO Leishmaniasis Control Team. Leishmaniasis worldwide and global estimates of its incidence. *PLoS ONE* **2012**, *7*, e35671. [CrossRef] [PubMed]
3. Alcover, M.M.; Rocamora, V.; Guillén, M.C.; Berenguer, D.; Cuadrado, M.; Riera, C.; Fisa, R. Case Report: Diffuse Cutaneous Leishmaniasis by *Leishmania infantum* in a Patient Undergoing Immunosuppressive Therapy: Risk Status in an Endemic Mediterranean Area. *Am. J. Trop. Med. Hyg.* **2018**, *98*, 1313–1316. [CrossRef] [PubMed]
4. Rajni, E.; Ghiya, B.C.; Singh, S.; Shankar, P.; Swami, T.; Jadon, D.S.; Negi, S.R.; Malik, M.; Khatri, P.K. Cutaneous leishmaniasis in Bikaner, India: Clinicoepidemiological profile; parasite identification using conventional, molecular methods and CL Detect$^{TM}$ rapid test, a new Food and Drug Administration-approved test. *Trop. Parasitol.* **2019**, *9*, 115–123. [CrossRef] [PubMed]
5. Olliaro, P.; Vaillant, M.; Arana, B.; Grogl, M.; Modabber, F.; Magill, A.; Lapujade, O.; Buffet, P.; Alvar, J. Methodology of Clinical Trials Aimed at Assessing Interventions for Cutaneous Leishmaniasis. *PLoS Negl. Trop. Dis.* **2013**, *7*, e2130. [CrossRef] [PubMed]
6. Jaafari, M.R.; Bavarsad, N.; Fazly-Bazzaz, B.S.; Samiei, A.; Soroush, D.; Ghorbani, S.; Lotfi-Heravi, M.M.; Khamesipour, A. Effect of Topical Liposomes Containing Paromomycin Sulfate in the Course of *Leishmania major* Infection in Susceptible BALB/c Mice. *Antimicrob. Agents Chemother.* **2009**, *53*, 2259–2265. [CrossRef]
7. Royal Society of Chemistry. Available online: https://doi.org/10.1039/9781788010177-00001 (accessed on 9 October 2019).
8. Trinconi, C.T.; Reimao, J.Q.; Bonano, V.I.; Espada, C.R.; Miguel, D.C.; Yokoyama-Yasunaka, J.K.U.; Uliana, S.R.B. Topical tamoxifen in the therapy of cutaneous leishmaniasis. *Parasitology* **2018**, *145*, 490–496. [CrossRef]
9. Alexandrino-Junior, F.; Silva, K.G.H.E.; Freire, M.C.L.C.; Lione, V.O.F.; Cardoso, E.A.; Marcelina, H.R.; Genre, J.; Oliveira, A.G.; Egito, E.S.T.D. A Functional Wound Dressing as a Potential Treatment for Cutaneous Leishmaniasis. *Pharmaceutics* **2019**, *11*, 200. [CrossRef]
10. Garnier, T.; Croft, S.L. Topical treatment for cutaneous leishmaniasis. *Curr. Opin. Investig. Drugs* **2002**, *3*, 538–544.
11. Zur, E. Topical Treatment of Cutaneous Leishmaniasis in Israel, Part 1. *Int. J. Pharm. Compd.* **2019**, *23*, 200–207.
12. El-On, J.; Livshin, R.; Even-Paz, Z.; Hamburger, D.; Weinrauch, L. Topical treatment of cutaneous leishmaniasis. *J. Investig. Dermatol.* **1986**, *87*, 284–288. [CrossRef] [PubMed]
13. Ben Salah, A.; Ben Messaoud, N.; Guedri, E.; Zaatour, A.; Ben Alaya, N.; Bettaieb, J.; Gharbi, A.; Belhadj Hamida, N.; Boukthir, A.; Chlif, S.; et al. Topical paromomycin with or without gentamicin for cutaneous leishmaniasis. *N. Engl. J. Med.* **2013**, *368*, 524–532. [CrossRef] [PubMed]
14. Sosa, N.; Pascale, J.M.; Jiménez, A.I.; Norwood, J.A.; Kreishman-Detrick, M.; Weina, P.J.; Lawrence, K.; McCarthy, W.F.; Adams, R.C.; Scott, C.; et al. Topical paromomycin for New World cutaneous leishmaniasis. *PLoS Negl. Trop. Dis.* **2019**, *13*, e0007253. [CrossRef] [PubMed]
15. Soto, J.; Fuya, P.; Herrera, R.; Berman, J. Topical paromomycin/methylbenzethonium chloride plus parenteral meglumine antimoniate as treatment for American cutaneous leishmaniasis: Controlled study. *Clin. Infect. Dis.* **1998**, *26*, 56–58. [CrossRef]
16. Carneiro, G.; Santos, D.C.; Oliveira, M.C.; Fernandes, A.P.; Ferreira, L.S.; Ramaldes, G.A.; Nunan, E.A.; Ferreira, L.A. Topical delivery and in vivo antileishmanial activity of paromomycin-loaded liposomes for treatment of cutaneous leishmaniasis. *J. Liposome Res.* **2010**, *20*, 16–23. [CrossRef]
17. Brugués, A.P.; Naveros, B.C.; Calpena Campmany, A.C.; Pastor, P.H.; Saladrigas, R.F.; Lizandra, C.R. Developing cutaneous applications of paromomycin entrapped in stimuli-sensitive block copolymer nanogel dispersions. *Nanomedicine (Lond.)* **2015**, *10*, 227–240. [CrossRef]
18. Frankenburg, S.; Glick, D.; Klaus, S.; Barenholz, Y. Efficacious topical treatment for murine cutaneous leishmaniasis with ethanolic formulations of amphoteicin B. *Antimicrob. Agents Chemother.* **1998**, *42*, 3092–3096. [CrossRef]

19. López, L.; Vélez, I.; Asela, C.; Cruz, C.; Alves, F.; Robledo, S.; Arana, B. A phase II study to evaluate the safety and efficacy of topical 3% amphotericin B cream (Anfoleish) for the treatment of uncomplicated cutaneous leishmaniasis in Colombia. *PLoS Negl. Trop. Dis.* **2018**, *12*, e0006653. [CrossRef]
20. Schmidt-Ott, R.; Klenner, T.; Overath, P.; Aebischer, T. Topical treatment with hexadecylphosphocholine (Miltex) efficiently reduces parasite burden in experimental cutaneous leishmaniasis. *Trans. R. Soc. Trop. Med. Hyg.* **1999**, *93*, 85–90. [CrossRef]
21. Dar, M.J.; Din, F.U.; Khan, G.M. Sodium stibogluconate loaded nano-deformable liposomes for topical treatment of leishmaniasis: Macrophage as a target cell. *Drug Deliv.* **2018**, *25*, 1595–1606. [CrossRef]
22. Moosavian Kalat, S.A.; Khamesipour, A.; Bavarsad, N.; Fallah, M.; Khashayarmanesh, Z.; Feizi, E.; Neghabi, K.; Abbasi, A.; Jaafari, M.R. Use of topical liposomes containing meglumine antimoniate (Glucantime) for the treatment of L. major lesion in BALB/c mice. *Exp. Parasitol.* **2014**, *143*, 5–10. [CrossRef] [PubMed]
23. Moosavian, S.A.; Fallah, M.; Jaafari, M.R. The activity of encapsulated meglumine antimoniate in stearylamine-bearing liposomes against cutaneous leishmaniasis in BALB/c mice. *Exp. Parasitol.* **2019**, *200*, 30–35. [CrossRef] [PubMed]
24. Anchisi, C.; Maccioni, A.M.; Sinico, C.; Valenti, D. Stability studies of new cosmetic formulations with vegetable extracts as functional agents. *Farmaco* **2001**, *56*, 427–431. [CrossRef]
25. Bilia, A.R.; Bergonzi, M.C.; Mazzi, G.; Vincieri, F.F. Development and stability of semisolid preparations based on a supercritical CO2 Arnica extract. *J. Pharm. Biomed. Anal.* **2006**, *41*, 449–454. [CrossRef] [PubMed]
26. Risaliti, L.; Piazzini, V.; Di Marzo, M.G.; Brunetti, L.; Cecchi, R.; Lencioni, P.; Bilia, A.R.; Bergonzi, M.C. Topical formulations of delta-aminolevulinic acid for the treatment of actinic keratosis: Characterization and efficacy evaluation. *Eur. J. Pharm. Sci.* **2018**, *115*, 345–351. [CrossRef] [PubMed]
27. Suñer, J.; Calpena, A.C.; Clares, B.; Cañadas, C.; Halbaut, L. Development of Clotrimazole Multiple W/O/W Emulsions as Vehicles for Drug Delivery: Effects of Additives on Emulsion Stability. *AAPS Pharm. Sci. Tech.* **2017**, *18*, 539–550. [CrossRef]
28. Council regulation (EC) no 440/2008 Test guideline for skin absorption: In vitro method (B45). *J. Eur. Union* **2008**, *142*, 438–443.
29. World medical association declaration of Helsinki: Ethical Principles for Medical Research Involving Human Subjects. *JAMA* **2013**, *310*, 2191–2194. [CrossRef]
30. Du Plessis, J.; Stefaniak, A.; Eloff, F.; John, S.; Agner, T.; Chou, T.C.; Nixon, R.; Steiner, M.; Franken, A.; Kudla, I.; et al. International guidelines for the in vivo assessment of skin properties in non-clinical settings: Part 2. Transepidermal water loss and skin hydration. *Skin. Res. Technol.* **2013**, *19*, 265–278. [CrossRef]
31. Pujol, A.; Urbán, P.; Riera, C.; Fisa, R.; Molina, I.; Salvador, F.; Estelrich, J.; Fernández-Busquets, X. Application of Quantum Dots to the Study of Liposome Targeting in Leishmaniasis and Malaria. *Int. J. Theoret. Appiel. Nanotech.* **2014**, *2*, 1–8. [CrossRef]
32. Carrió, J.; de Colmenares, M.; Riera, C.; Gállego, M.; Arboix, M.; Portús, M. Leishmania infantum: Stage-specific activity of pentavalent antimony related with the assay conditions. *Exp. Parasitol.* **2000**, *95*, 209–214. [CrossRef] [PubMed]
33. Carrió, J.; Riera, C.; Gállego, M.; Portús, M. In vitro activity of pentavalent antimony derivates on promastigotes and intracellular amastigotes of Leishmania infantum strains from humans and dogs in Spain. *Acta Trop.* **2001**, *79*, 179–183. [CrossRef]
34. Handler, M.Z.; Patel, P.A.; Kapila, R.; Al-Qubati, Y.; Schwarts, R.A. Cutaneous and mucocutaneous leishmaniasis: Differential diagnosis, diagnosis, histopathology, and management. *J. Am. Acad Dermatol.* **2015**, *73*, 911–926. [CrossRef] [PubMed]
35. Dong, L.; Liu, C.; Cun, D.; Fang, L. The effect of rheological behavior and microstructure of the emulgels on the release and permeation profiles of Terpinen-4-ol. *Eur. J. Pharm. Sci.* **2015**, *78*, 140–150. [CrossRef]
36. Sosa, L.; Calpena, A.C.; Silva-Abreu, M.; Espinoza, L.C.; Rincón, M.; Bozal, N.; Domenech, O.; Rodríguez-Lagunas, M.J.; Clares, B. Thermoreversible Gel-Loaded Amphotericin B for the treatment of Dermal and Vaginal Candidiasis. *Pharmaceutics* **2019**, *11*, 312. [CrossRef]
37. Kaur, L.; Jain, S.K.; Singh, K. Vitamin E TPGS based nanogel for the skin targeting of high molecular weight anti-fungal drug: Development and *in vitro* and *in vivo* assessment. *RSC Adv.* **2015**, *5*, 53671–53686. [CrossRef]
38. Elias, P.M. Epidermal lipids, barrier function, and desquamation. *J. Investig. Dermatol.* **1983**, *80*, 44–49. [CrossRef]

39. Marzulli, F.N.; Callahan, J.F.; Brown, D.W. Chemical structure and skin penetrating capacity of a short series of organic phosphates and phosphoric acid. *J. Investig. Dermatol.* **1965**, *44*, 339–344. [CrossRef]
40. Anderson, B.D.; Raykar, P.V. Solute structure-permeability relationships in human stratum corneum. *J. Investig. Dermatol.* **1989**, *93*, 280–286. [CrossRef]
41. Rins, M.; Diez, I.; Calpena, A.C.; Obach, R. Skin density in the hairless rat. Evidence of regional differences. *Eur. J. Drug Metab. Pharmacokinet.* **1991**, *3*, 456–457.

pharmaceutics

MDPI

Article

# Development of Amphotericin B Micellar Formulations Based on Copolymers of Poly(ethylene glycol) and Poly(ε-caprolactone) Conjugated with Retinol

**Yeimy J. Rodriguez [1], Luis F. Quejada [2], Jean C. Villamil [2], Yolima Baena [3,*], Claudia M. Parra-Giraldo [2,*] and Leon D. Perez [1,*]**

1 Grupo de Investigación en Macromoléculas, Departamento de Química, Facultad de Ciencias, Universidad Nacional de Colombia-Sede Bogotá, Carrera 45 No. 26-85, Edificio 451 of. 449, Bogotá D.C. 11001, Colombia; yjrodriguezmo@unal.edu.co

2 Unidad de Proteómica y Micosis Humanas, Grupo de Enfermedades Infecciosas Departamento de Microbiología, Facultad de Ciencias, Pontificia Universidad Javeriana, Carrera 7 No. 43-82, Bogotá D.C. 110231, Colombia; lquejada@javeriana.edu.co (L.F.Q.); jcvillamilp@unal.edu.co (J.C.V.)

3 Grupo de Investigación SILICOMOBA, Departamento de Farmacia, Facultad de Ciencias, Universidad Nacional de Colombia-Sede Bogotá, Carrera 45 No. 26-85, Edificio 451 of. 449, Bogotá D.C. 11001, Colombia

* Correspondence: ybaenaa@unal.edu.co (Y.B.); claudia.parra@javeriana.edu.co (C.M.P.-G.); ldperezp@unal.edu.co (L.D.P.); Tel.: +57-1-3165000 (Y.B.); +57-1-3208320 (C.M.P.-G.); +57-1-3165000 (L.D.P.)

Received: 23 October 2019; Accepted: 22 January 2020; Published: 25 February 2020

**Abstract:** Amphotericin B (AmB) is a broad spectrum of antifungal drug used to treat antifungal diseases. However, due to the high toxicity of AmB, treated patients may suffer the risk of side effects, such as renal failure. Nanoencapsulation strategies have been reported to elicit low toxicity, albeit most of them possess low encapsulation efficiency. The aim of this research is to develop micellar delivery systems for AmB with reduced toxicity while maintaining its affectivity by employing retinol (RET)-conjugated amphiphilic block copolymers (ABCs) as precursors. Copolymers composed of poly(ε-caprolactone) (A) and polyethylenglycol (B) of types AB and ABA were synthesized by ring opening polymerization and subsequently conjugated with RET by Steglich esterification. $^1$H-NMR spectroscopy was used to corroborate the structure of copolymers and their conjugates and determine their molecular weights. Analysis by gel permeation chromatography also found that the materials have narrow distributions. The resulting copolymers were used as precursors for delivery systems of AmB, thus reducing its aggregation and consequently causing a low haemolytic effect. Upon conjugation with RET, the encapsulation capacity was enhanced from approximately 2 wt % for AB and ABA copolymers to 10 wt %. AmB encapsulated in polymer micelles presented improved antifungal efficiency against *Candida albicans* and *Candida auris* strains compared with Fungizone®, as deduced from the low minimum inhibitory concentration.

**Keywords:** Amphotericin B; polymer micelles; drug delivery

## 1. Introduction

During the last decades, the number of people susceptible to the invasive fungal disease (IFD) has increased. It is a consequence of advances in medicine that have increased the life expectancy of HIV-infected patients and augmented the number of procedures, such as organ and bone marrow transplants. Furthermore, the population growth also augments the occurrence of premature births and diseases (i.e., cancer) that can compromise the immune system [1,2]. Accelerated environmental deterioration, extreme weather conditions and limited access to the health system also contribute to the incidence [3]. The diagnosis and treatment of IFD is difficult in many cases due to the minimal specificity of its symptomatology, especially since the identification of the pathogen source has further become complex, mainly in hospital centres where technological resources are unavailable or do not employ a personnel trained for this purpose; resorting to empirical therapies, which implies the use of broad spectrum antifungal drugs [4].

Since 1959, Amphotericin B (AmB) has been the "gold standard" for the treatment of IFDs. Fungizone® (a colloidal dispersion of this drug is stabilized by sodium deoxycholate as a surfactant) is the most widely used formulation of AmB due to its effectiveness, wide spectrum of action and low number of reports of resistant strains [5–7]. However, the toxicity of AmB and its bioaccumulation in organs, such as liver, lungs and kidneys, are the undesirable features that seriously affect the treated patients [8,9]. According to a study by Wingard et al. [10], the probability of patients receiving the above medication to suffer from kidney damage could be so high as 70%, indicating both an increase in mortality rate and cost of treatment [11]. With the development of liposomal formulations, among them Ambisome®, the toxicity of AmB has been attenuated due to the small size of the liposomes that allows for a prolonged circulation and distribution into the organs. Additionally, the strong interaction between AmB and the liposome components endow a controlled release to the drug [12]. However, high cost apart from high dose (5–10 mg/kg for Ambisome® compared with 0.5–1 mg/kg for Fungizone®) requirements are disadvantageous, principally among low- and middle-income countries, such as Colombia [13–17].

Although AmB has been used for 60 years, a debate persists about the mechanism for its antifungal action. The firstly and most accepted model is the formation of barrel-type pores that permeate the double-layer fungus cell membrane. According to this model, the heptaene segments are oriented towards the outside of the pore, while the hydroxyl groups are arranged towards the inside, allowing the cell to lose ions and polar molecules which are vital for its functioning. Pore formation is mediated by the complexation of membrane sterols with AmB molecules [18]. Other mechanisms, such as sequestering of sterols by aggregates of AmB on the surface of cells (sterol sponge) [19], and the formation of reactive nitrogen and oxygen species promoted by AmB that cause irreversible damages to cell integrity, has also being proposed [9].

AmB, despite being an active antifungal drug, is a highly cytotoxic substance, particularly against human red blood cells (RBCs) [20]. This fact is also attributed to its ability to induce cholesterol complexation, which is a component of cell membranes. AmB has a structure resembling ergosterol and therefore has a similar binding pathway. According to simulations by molecular dynamics and density functional theory, the interaction of AmB with sterol occurs through van der Waals forces, $\pi$–$\pi$ interactions and hydrogen bonds [21]. However, an experimental probe found that the interaction of AmB with ergosterol is stronger than that with cholesterol [22]. The preference of AmB for ergosterol, with their structural similarities, is attributed to the high degree of unsaturation, especially to conjugate ring B, which increases the magnitude of $\pi$–$\pi$ interactions with heptane segments of AmB and the van der Waals forces that depend on molecular geometries. A research also suggests that complex formation is less entropically favoured given the high structural flexibility of cholesterol [23].

AmB, with its amphiphilic nature, aggregates when its concentration exceeds its critical aggregation concentration (CAC) of approximately 1 μg/mL [24]. Past research proved that AmB, which occurs in monomeric form in low concentrations, has a high tendency to interact with ergosterol-rich membranes, but cholesterol-containing membranes are unaffected [25]. However, AmB also aggregates and acts as sponge to sequester cholesterol [24]. The presence of AmB water-soluble aggregates induce haemolytic effects on human and mouse erythrocytes; meanwhile, the formulations dominated by the monomeric form have low cytotoxicity [26].

Nowadays, polymer micelles (PMs) have emerged as one the main carriers in the development of nanostructured drug formulations [27,28]. These nanoparticles are composed of a hydrophobic core which is capable of encapsulating water-insoluble drugs such as AmB, and a hydrophilic shell that confers colloidal stability and compatibility with biological systems [29,30]. A recent report indicates that upon administration, PMs dissociate, due to their dilution and interaction with plasmatic constituents such as proteins, producing individual block copolymer chains which are quickly sequestered by Kupffer cells into the liver [31]. It suggests that, in addition to the controlled release of the cargos, PMs are safe nanocontainers for nephrotoxic drugs such as AmB. This is in concordance with in vivo studies which corroborate that micellar systems composed of polystyrene-block-polyethylene oxide [32], linoleic acid modified polyethylene glycol-block-polyethylenimine [33], polyethylene glycol-block-poly(*N*-hexyl stearate l-aspartamide) [34] and partially benzylated poly-L-aspartic acid [35] endow AmB lower toxicity.

The aim of this study is to obtain micellar vehicles for AmB with improved encapsulation efficiency, control the release of the drug and lessen its cytotoxic effect without altering its antifungal effectivity. Amphiphilic diblock and triblock copolymers composed of segments of poly(ethylene glycol) (PEG) and poly(ε-caprolactone) (PCL), which are biodegradable and biocompatible polymers and widely used in biomedical applications, were initially synthesized [36–38]. These materials were subsequently conjugated with retinol (RET), which is a hydrophobic biomolecule with a polyenic segment, and evaluated as vehicles for AmB. The resulting formulations were characterized to determine the aggregation state of the drug, its release kinetics, cytotoxicity and antifungal activity.

## 2. Materials and Methods

### 2.1. Materials

ε-Caprolactone (CL) 98%, methoxy-polyethylene glycol (mPEG) of 5 kDa and polyethylene glycol diol (PEG-diol) of 6 kDa, RET (95%), pyrene (99%), AmB (98%), 4-(dimethylamino)pyridine (DMAP) (99%), Tin(II) 2-ethylhexanoate ($Sn(Oct)_2$) (95%), succinic anhydride (≥99%), *N,N'*-dicyclohexylcarbodiimide (DCC), triethylamine (TEA), resazurin sodium salt, ethylenediaminetetraacetic acid (EDTA), Methylthiazolyldiphenyl-tetrazolium bromide (MTT), Dulbecco's Modified Eagle's Medium (DMEN) and other reagents and solvents used in the protocols of synthesis, purification and characterization were purchased from Sigma-Aldrich, St. Luis, MO, USA. Fungizone® manufactured by Abbott was provided by a local supplier. Before any synthesis procedure was conducted, toluene and tetrahydrofuran (THF) were dried by reactive distillation with sodium. mPEG and PEG-diol were dried by azeotropic distillation with dried toluene. Traces of water in CL were removed using calcium hydride ($CaH_2$) for 48 h and stored with a 3 Å molecular sieve. The other reagents were used without any additional purification.

### 2.2. Synthesis Procedures

#### 2.2.1. Synthesis of PEG-b-PCL Copolymers

In this study, ABCs of types AB and ABA composed of polycaprolactone (A) and polyethylene glycol (B) were synthesized by ring opening polymerization. Firstly, CL was polymerized by ring opening polymerization, starting from PEG as the initiator and $Sn(Oct)_2$ as the catalyst. In typical synthesis conditions, mPEG (0.6 mmol) as the initiator and $Sn(Oct)_2$ (0.6 mmol) as the catalyst are dissolved in dried toluene under Ar atmosphere, followed by CL (7.9 mmol). Here, the reaction mixture was maintained under stirring at 115 °C for 24 h. The ABA copolymer was synthesized in the same conditions but starting from PEG-diol as the initiator. The polymers were purified by passing them through a column packed with basic alumina and by three solvent precipitation cycles. The solid product was dried under reduced pressure at room temperature. In $^1H$-NMR (400 MHz, $CDCl_3$, δ, ppm), the distinctive signals of PCL were as follows: 4.08 (t, J = 6.6), 2.33 (t, J = 6.6 Hz, 73H), 1.72–1.62 (m, 146H), 1.40 (m, 73H) and mPEG: 3.66 (s, 494H). The spectrum is shown in Figure S1A.

2.2.2. Synthesis of PEG-PCL-COOH

AB and ABA copolymers were converted into their corresponding carboxylated materials by esterification with succinic anhydride following the steps of a previous procedure [39]. Briefly, the corresponding hydroxyl ended copolymer (5 mmol) and SA (6 mmol) were dissolved in 10 mL of chloroform. After adding DMAP (6 mmol) and TEA (6 mmol), the reaction mixture was maintained under stirring for 24 h in Ar atmosphere. In recovering the product, the reaction mixture was firstly filtrated, and then the modified polymer was precipitated by adding an excess of diethyl ether. The carboxylated copolymer was purified in two successive precipitations. Finally, the recovered solid was dried under vacuum at room temperature. In the $^1$H-NMR spectrum, apart from signals for the PEG and PCL segments, a new signal at 2.66 ppm due to methylene groups of succinic moiety also appeared. The spectrum is provided in Figure S1B.

2.2.3. Synthesis of PEG-PCL-RET

Carboxylated copolymers were reacted with RET by the Steglich esterification procedure. The materials (2 mmol of –COOH) and DCC (1 mmol) were dissolved in 15 mL of DCM and stirred at 500 rpm for 30 min. Then, DMAP (6 mmol) and RET (5 mmol) were added. The reaction mixture was stirred at room temperature in darkness for 24 h [40]. Finally, the reaction mixture was filtered to remove the by-product *N,N′*-dicyclohexylurea. The RET-conjugated AB and ABA copolymers were purified by solvent precipitation with DCM and ethyl ether. Until the usage of these copolymers, they were stored at 6 °C under inert atmosphere.

*2.3. Characterization Techniques*

The $^1$H-NMR spectra of mPEG and HO-PEG-OH, AB and ABA copolymers and their corresponding carboxylated and RET-conjugated derivatives were recorded on a 400 Ultrashield spectrometer operated a 400 MHz (Bruker, Mannheim, Germany) by using CDCl3 as the solvent. The FTIR spectra were acquired in an IR Prestige 21 FTIR equipment (Shimadzu, Houston, TX, USA). The samples were dissolved in THF and deposited on NaCl window. After solvent evaporation, the resulting film was characterized by transmittance. Molecular weight distribution and dispersion index were determined by gel permeation chromatography using a high-performance liquid chromatographer (Waters, Pittsburgh, PA, USA) equipped with a differential refraction index detector. Analyses were performed in THF at a flow rate of 0.8 mL/min by using HR 4E column. The calibration curve was constructed with polystyrene standards in the range of 1–69 kDa.

The thermal behaviour of the polymeric materials was characterized by a differential scanning calorimeter (DSC, Model 822e, Mettler Toledo, Colombus, OH, USA) as follows. Firstly, thermal history was erased by heating from room temperature to 100 °C at 20 °C/min. Then, the samples were cooled down to −60 °C at 10 °C/min to allow for crystallization. Finally, the melting behaviour was determined by heating them to 250 °C at 10 °C/min.

The average size and distribution of particles were determined by dynamic light scattering (DLS, Litesizer 500, Anton Paar, Houston, TX, USA) operated at 27 °C and 37 °C. The samples were diluted with deionized water (approximately 18 MOhm-cm) to a concentration of 0.1%.

*2.4. Critical Micellar Concentration Measurements*

The critical micellar concentration (CMC) of block copolymers was determined by fluorescence and using pyrene as a probe based on previously reported procedures [41–43]. Briefly, micellar dispersion of AB, ABA and their conjugates were prepared in various concentrations (0.01 to 1000 mg/L) with a fixed concentration of pyrene of 0.6 μM. The excitation spectra of pyrene from 300 to 360 nm were monitored at 390 nm for each dilution by using a Cary Eclipse fluoresce spectrometer (Agilent, Santa Clara, CA, USA). CMC was determined by plotting the ratio of the intensities at 335 and 332 nm ($I_{335}/I_{332}$) versos log concentrations.

### 2.5. Preparation of AmB Loaded Micelles

AmB@PMs was prepared in the modified nanoprecipitation method [43]. Briefly, 20-mg of the corresponding polymeric sample was completely dissolved in THF, and the resulting solution was dropped to 25.0 mL of deionized water at 25 °C. Simultaneously, 5 mg of AmB was dissolved in approximately 5 mL of methanol, and this solution was set to dropwise relative to the solution containing the polymer. The final mixture was maintained under continuous stirring for 48 h. After the complete evaporation of the solvent achieved under reduced pressure, non-encapsulated AmB was removed by centrifuging the medicated emulsions at $1.3 \times 10^4$ RPM for 14 min. AmB@PM suspensions were dried by lyophilisation and maintained at −20 °C until the biological and physical characterizations were completed.

### 2.6. Determination of Encapsulated AmB

Encapsulated AmB was determined as follows. Approximately 1 mg of the corresponding formulation was dispersed in approximately 8 mL of methanol. This mixture was sonicated for 15 min at room temperature, and then the volume was completed to 10.00 mL. This mixture was centrifuged at 5000 RPM for 10 min to remove the polymer component. The supernatant was analysed by UV-vis at 406 nm. The measurement allowed for the determination of drug loading (%DL) and encapsulation efficiency (%EE) with Equations (1) and (2), respectively.

$$\%DL = \frac{mass\ of\ encapsulated\ AmB}{mass\ of\ AmB@PMs} \times 100 \quad (1)$$

$$\%EE = \frac{mass\ of\ encapsulated\ AmB}{mass\ of\ feeded\ AmB} \times 100 \quad (2)$$

### 2.7. Assessment of Aggregation State

The aggregation degree of AmB in the polymer micelles was studied by UV-vis spectrometry. The spectra of the solutions of AmB in DMSO and PBS, and the dispersions of AmB@PMs in PBS at pH 7.4, were directly acquired at 300–450 nm in an Evolution 300 UV/VIS spectrometer. All the solutions were prepared to a final concentration of AmB of 3.8 μg/mL.

### 2.8. Release Study

The release kinetics of AmB from formulated polymer micelles was assessed in vitro by employing the dialysis method [44]. A mass of each formulation corresponding to approximately 0.25 mg of AmB was dispersed in approximately 3 mL of PBS and then placed in a cylindrical tube capped with a dialysis membrane and MWCO of 10 KDa. The formulation was dialyzed against 40.0 mL of a solution of sodium deoxycholate (0.5%) and DMSO in the volume ratio of 2:1. The release was measured for 96 h at 37 °C. The profiles were obtained by periodically withdrawing aliquots of 1.00 mL. The measurement of the released AmB was performed by UV–vis.

The volume of the release medium was maintained constant by repositioning each drawn aliquot with fresh medium. The cumulative release ($Q$) was calculated as follows:

$$Q = C_n V + \sum_{i=1}^{i=n-1} C_i V_i, \quad (3)$$

where $V$ is the volume of the release medium, and $C_n$ and $C_i$ correspond to concentration at a given time and the former aliquots with volume $V_i$, respectively.

## 2.9. In Vitro Biocompatibility

### 2.9.1. Haemolysis

Analysis was performed following a published protocol [44]. Whole blood collected from O+ donors was added with an anticoagulant 1 mM of EDTA solution and then centrifuged at 500× *g* for 5 min to separate the RBCs. The recovered RBCs were rinsed twice with PBS and diluted in the same buffer to obtain an absorbance of 0.5 at 540 nm. Then, 190 µL of this cell dispersion was treated with 10 µL of a solution containing the corresponding testing material. PBS was used as a negative control, whereas in the presence of Triton X-100 complete haemolysis of RBCs is guaranteed. The treated solutions of RBCs were incubated at 37 °C under continuous shaking for 1 h. Afterward, the plaque was centrifuged at 500× *g* for 5 min to separate the non-lysed RBCs. The haemolysis was determined by measuring the absorbance of the haemoglobin in the solution at 541 nm as follows:

$$Hemolysis(\%) = \frac{A_s - A_{NC}}{A_{PC} - A_{NC}} \times 100, \quad (4)$$

where $A_s$ is the absorbance of the drug-treated RBCs, and $A_{NC}$ and $A_{PC}$ are the absorbance values of the negative and positive controls, respectively.

### 2.9.2. Cytotoxicity against Fibroblasts

The cytotoxicity of AmB@PMs was evaluated on mouse fibroblast (L929) using the MTT colorimetric assay following a previously published protocol [44]. Briefly, fibroblasts were seeded at a density of $2.1 \times 10^4$ cells/well, allowed to adhere and proliferate for 24 h in DMEN medium. Wells containing the adhered cells were treated with dispersions of AmB@PMs corresponding to concentrations of AmB of 0.05 to 15 mg/L. The cell cultures were incubated along 24 h at 37 °C under a 5% $CO_2$ atmosphere. Finally, the medium was replaced with 30 µL of a solution of 1 mg/L of MTT in PBS. The treated cells were incubated at 37 °C for 4 h, then MTT was removed. The resulting Formazan crystals were solubilized with 100 µL, and analysed in an ELx800 reader (BioTek, Winooski, VT, USA) at 490 nm. The absorbance values were computed as indicated in Equation (5) to determine cell viabilities.

$$Cell\ Viability\ (\%) = \frac{A_{Sample}}{A_{Blank}} \times 100, \quad (5)$$

where $A_{sample}$ is the absorbance of AmB@PMs and Fungizone® treated well, and $A_{Blank}$ is the absorbance of control well without drug treatment.

## 2.10. In Vitro Assessment of Antifungal Activity

An antifungal susceptibility test was carried in accordance with Clinical and Laboratory Standards Institute's (CLSI) broth microdilution method (BMD) following the M27-A3 guidelines [45,46]. Dilutions of AmB@PMs and Fungizone® with concentrations of AmB in the range of 0.11–7.5 µg/mL were evaluated. Minimum Inhibitory Concentrations (MICs) were determined visually after 24 h as the lowest concentration of drug that can cause considerable diminution relative to the drug-free growth control. The MICs were determined visually and supported by measurements of absorbance, which was determined by spectrophotometry at 600 nm. The values obtained at this phase were further corroborated employing resazurin (7 mM) as a redox indicator. Resazurin is a blue non-toxic cell-permeable compound and virtually non-fluorescent. Upon entering living cells, resazurin is reduced to resorufin, a compound that is violet to pink and highly fluorescent [47,48].

## 3. Results

### 3.1. Synthesis of Polymeric Precursors

mPEG-b-PCL (AB) and PLC-b-PEG-b-PCL (ABA) block copolymers were synthesized by Ring Opening Polymerization (ROP) starting from PEG as the macroinitiator (Scheme 1A). Characterizations by $^1$H-NMR confirmed that the proposed structure was achieved. The average molecular weights (Mn) are listed in Table 1. The dispersion indexes estimated by gel permeation chromatography coincide with values reported for polymerization catalysed by Sn(Oct)$_2$ [49].

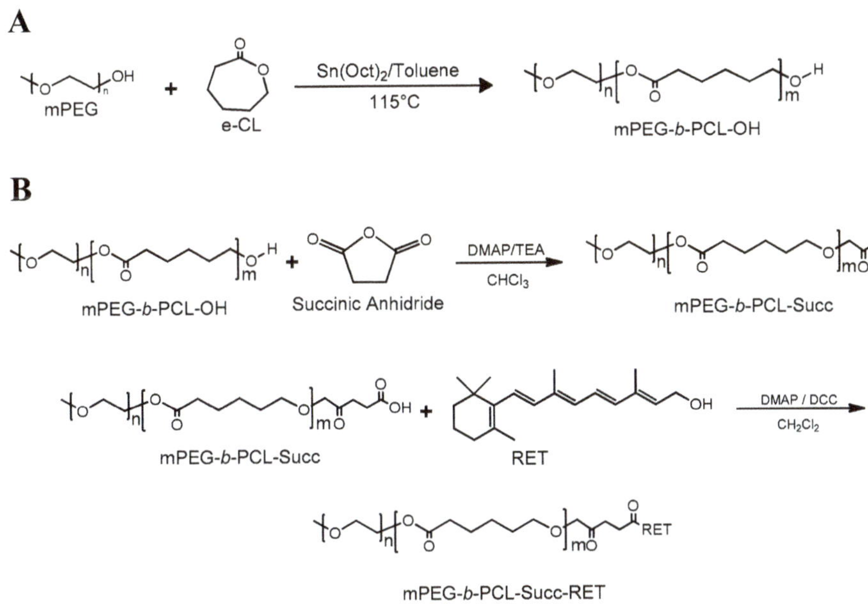


**Scheme 1.** Synthesis pathway. (**A**) Ring Opening Polymerization (ROP) of CL by using mPEG as the initiator, and (**B**) conjugation of mPEG-b-PCL with RET by Steglich esterification.

**Table 1.** Characterization of mPEG-b-PCL (AB) and PLC-b-PEG-b-PCL (ABA) and their corresponding conjugated structures.

| Sample | Composition [a] | $M_n$ (kDa) [a] | $M_w/M_n$ [b] | CMC | |
|---|---|---|---|---|---|
| | | | | mg/L | µM |
| AB | (mPEG)$_{114}$-(PCL)$_{11}$-OH | 6.27 | 1.20 | 4.5 ± 0.26 | 4.9 |
| AB-RET | (mPEG)$_{114}$-b-(PCL)$_{11}$-RET | 6.56 | 1.24 | 7.4 ± 0.78 | 8.0 |
| ABA | HO-(PCL)$_9$-(PEG)$_{136}$-(PCL)$_9$-OH | 8.04 | 1.13 | 2.7 ± 0.86 | 2.9 |
| ABA-RET | RET-(PCL)$_9$-(PEG)$_{136}$-(PCL)$_9$-RET | 8.61 | 1.29 | 5.5 ± 0.72 | 5.9 |

[a] Determined by integration of $^1$H-NMR signals at 3.66 pm assigned to methylene groups of PEG which was taken as a molecular weight reference and –CH$_2$O groups of PCL at 4.08 ppm, and [b] measured by GPC by employing a relative calibration plot.

Afterward, the resulting AB and ABA copolymers were reacted with succinic anhydride to obtain a carboxylate copolymer at the PCL ends. The occurrence of this reaction was confirmed by the appearance of a new signal near 2.7 ppm due to succinic acid end moieties (Figure S1). The carboxylated materials were subsequently esterified with RET by the Steglich process in the presence of DCC and DMAP as the coupling agent and the catalyst, respectively [50] (Scheme 1B). The occurrence of this reaction was confirmed by $^{1}$H-NMR and $^{13}$C-NMR. The corresponding spectra of the isolated products (Figure 1A,B) have signals resulting from PEG, PCL and RET moieties. 2D-NMR spectra were not feasible given the low number of RET nuclei compared to the copolymer.

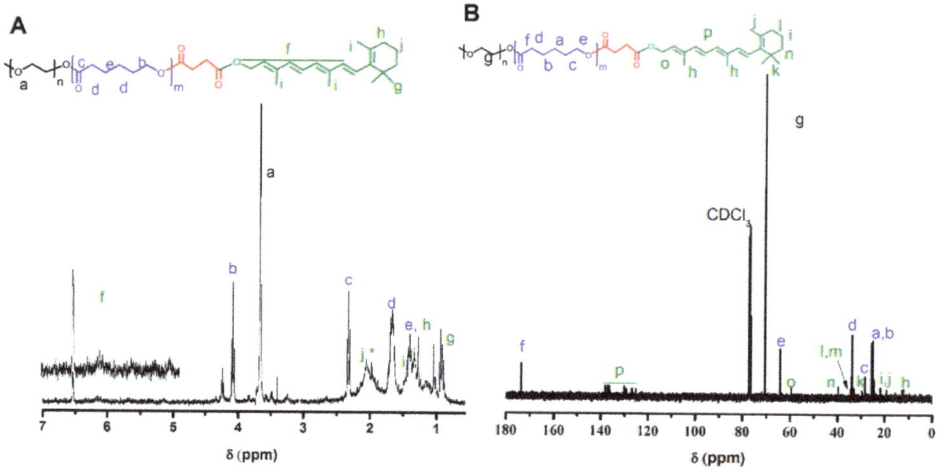


**Figure 1.** (**A**) $^{1}$H and (**B**) $^{13}$C-NMR spectra of mPEG-b-PCL conjugated with retinol and their corresponding signal designation.

## *3.2. Critical Micellar Concentration*

In aqueous media, ABCs, as a result of solvophobic forces, assemble themselves to produce aggregates; among them, micelles are more frequently reported for low molecular weight copolymers [30]. This kind of core–shell structure is spontaneously formed when the concentration of ABC is larger than its CMC. Thus, determining the CMC allows for the characterization of the thermodynamic favourability of the self-assembling process, as well as the stability of the resulting nanostructures against dilution. Additional to thermodynamic stability, PMs even at concentrations below the CMC of the corresponding ABC remain stable for long periods as a result of intermolecular interaction in polymer segments [51]. Conversely, in the presence of a drug the stability of the PMs (not studied here) is altered depending on the nature of the polymer–cargo interactions [52].

In this study, CMC values were measured by fluorimetry employing pyrene as a probe molecule, this technique apart from being a simple method, has also shown to be sensitive to obtain reliable values [43,53,54]. Figure 2A shows a spectral comparison of pyrene dissolved in solutions with different concentration of representative copolymers. Apart from the obvious increase in fluorescence intensity when the concentration increases, a red shift of the maxima is also observed. Figure 2B presents a plot of the ratio of intensities at 335 and 332 nm. At the low copolymer concentration, this ratio is kept constant. However, the ratio increases at certain concentration. This phenomenon reveals the presence of micelles and that CMC has been achieved, thus allowing for the measurement of the values listed in Table 1. According to these results, the AB copolymer exhibits a larger CMC than ABA, suggesting a more hydrophilic character. In conjugation with RET, the CMC increases, showing a tendency opposed to the highly hydrophobic character of the PCL segment that was achieved in the presence of RET. This finding suggests that the conjugation of AB and ABA decreases segmental flexibility, corresponding to a relatively high steric hindrance that increase of energy required to achieve conformational changes required for the micelles formations.

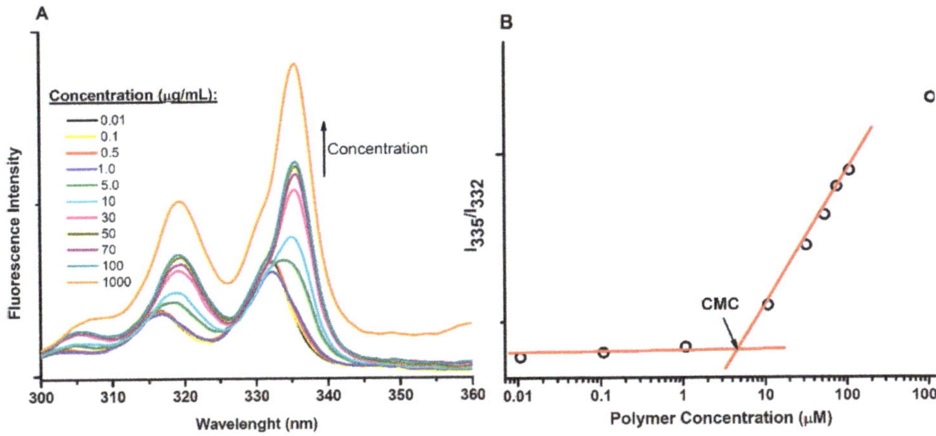


**Figure 2.** CMC measurement by fluorescence employing pyrene as the probe molecule. (**A**) Emission spectra of solutions containing different concentration of a representative sample. (**B**) Plot of $I_{335}/I_{332}$ used to measure CMC values.

### 3.3. *Differential Scanning Calorimetry*

DSC was used to study the effect of the conjugation of AB and ABA copolymers with RET on their thermal properties (i.e., their morphology). After erasing the thermal history of the samples, the crystallization of the samples was studied by cooling them down from 100 to −20 °C at 10 °C/min, while heating the samples to as high 100 °C at 10 °C/min allowed for the characterization of their melting.

According to Figure 3A showing representative traces of ABA, PEG-diol and ABA-RET, the samples exhibit a crystallization transition characterized by the release of thermal energy corresponding to crystallization enthalpy. Table 2 lists the temperatures that correspond to the maxima of the crystallization peaks. According to these values, PEG homopolymers crystallize at a higher temperature than the corresponding copolymers, proving that $CH_2$–O–$CH_2$ linkages endow PEG flexibility and allows for the chain folding required to achieve high crystallization. Upon copolymerization with PCL to obtain AB and ABA copolymers, the crystallization temperature is shifted to a low temperature, indicating that this transition is kinetically less favoured. Cooling traces of AB support the occurrence of two separated crystallization phenomena, while ABA exhibited a single peak with a shoulder at the high temperature corresponding to the superposition of two transitions; these findings suggest the formation of segregated crystalline domains [55,56]. Moreover, in conjugation with RET, the crystallization was delayed, as deduced from the shift to the low temperature, and the fact that transition became less pronounced, as seen for ABA-RET in Figure 3B. The presence of RET reduced the crystallinity of both segments by reducing their segmental mobility.

Figure 3B shows the DSC traces obtained in the heating stage for the representative materials. PEG-diol exhibits two melting peaks characterizing different crystalline forms of PEG [57]. However, ABA only presents a broad peak with a maximum at 54.4 °C that can be attributed to the melting of both PEG and PCL domains. The depletion of $T_m$ of the mPEG segment agrees with a low intense crystallization peak, low crystallization degree and a decrease of size of crystalline domains. When ABA was conjugated with RET, besides the notable decrease in $T_m$, the appearance of a cool crystallization ($T_{cc}$) peak at approximately −10 °C corroborates that RET hinders the crystallization of both PEG and PCL domains. The data shown in Table 2 indicate the same tendency for the diblock materials. Meanwhile, melting enthalpy ($\Delta H_m$), which was estimated by the integration of melting peaks, corroborates the decrease in crystallization degree upon the conjugation with RET; this phenomenon is in concordance with the reduction of segmental mobility.

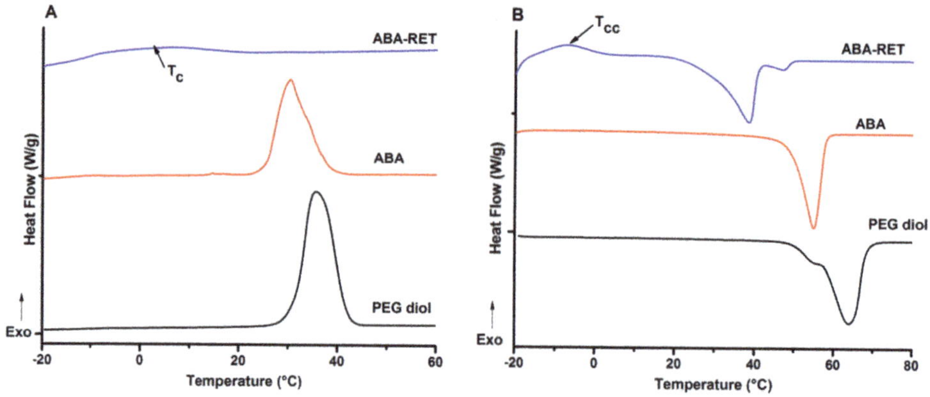


**Figure 3.** DSC traces for a representative set of materials: PEG-diol, PLC-b-PEG-b-PCL (ABA and its corresponding conjugate ABA-RET). Cooling (**A**) and heating (**B**).

**Table 2.** Thermal properties of the polymeric materials measured by differential scanning calorimeter (DSC).

| Sample | Thermal Properties | | |
|---|---|---|---|
| | $T_c$ (°C) | $T_m$ (°C) | $\Delta H_m$ (J/g) |
| mPEG | 40.6 | 60.2 | 171.9 |
| AB | 28.2<br>18.6 | 48.0 | 86.2 |
| AB-RET | 23.3<br>−7.15 | 39.4 | 61.9 |
| PEG-diol | 35.4 | 63.9 | 107.9 |
| ABA | 30.8 | 54.4 | 110.7 |
| ABA-RET | 2.7 | 38.41<br>49.1 | 41.8 |

### 3.4. AmB Encapsulation

The encapsulation of AmB in polymer micelles was carried out by a partition method. A colloidal dispersion of the preformed micelles was added with AmB dissolved in an excess of methanol. This setup allowed for the diffusion of AmB from the continuous medium to the micelle core. When methanol evaporates, AmB can self-assemble in an aqueous medium to produce insoluble aggregates, or it can interact with the components of micelles driven to its encapsulation. Table 3 contains the results of the quantification of encapsulated AmB carried out on the lyophilized formulations after extracting the AmB with methanol. The analysis of its concentration was performed with UV-vis spectroscopy to obtain the provided DL and EE values.

**Table 3.** Drug loading (DL) and encapsulation efficiency (EE) obtained by analysis by UV-vis the lyophilized formulations.

| Formulation | EE (%) | DL (%) |
|---|---|---|
| AmB@AB | 13.7 ± 0.71 | 2.74 ± 0.17 |
| AmB@AB-RET | 51.05 ± 0.79 | 10.21 ± 0.21 |
| AmB@ABA | 10.19 ± 0.12 | 2.38 ± 0.03 |
| AmB@ABA-RET | 33.65 ± 1.23 | 6.73 ± 0.28 |

The results indicate that encapsulation efficiency and loading capacity increase in the presence of RET, suggesting favourable polymer–AmB chemical interactions as corroborated by the most pronounced increase achieved for ABA-RET. Both RET and AmB presented a polyenic segment that likely favours π–π interactions [39]. This tendency agrees well with previous findings that corroborate an increase in encapsulation efficiency when PEG-b-PCL copolymers are conjugated with cholesterol [44]. Similarly, the conjugation of AB and ABA copolymers with RET decreases their crystallinity, thereby increasing the free volume of the polymer chains, which in turn favours the encapsulation of the drugs.

Both empty micelles and micelles loaded with AmB were characterized by dynamic light scattering at 25 °C and 37 °C to determine their sizes and distributions (Table 4). Empty PMs composed of RET-conjugated copolymers presented a larger diameter than the corresponding initial material, indicating that the structure of the micelles and the nature of the interactions leading to their formation depend on the structure of the copolymers. A3s the temperature is increased, the diameter of the particles also increases, suggesting a reduction of the solvation of the hydrophilic PEG segment. Thus, the aggregation of particles is enabled [58]. Likewise, a temperature augment could also cause a larger aggregation number as well as a change of the polymer chains conformations [59,60]. According to DLS results, no evidence about AmB leakage upon heating the colloidal dispersion was obtained.

**Table 4.** Average diameters measured by dyanimc light scattering (DLS) for empty micelles and AmB@PMs.

| Formulation | Empty Micelles | | | | AmB@PMs | | | |
|---|---|---|---|---|---|---|---|---|
| | 25 °C | | 37 °C | | 25 °C | | 37 °C | |
| | Dh (nm) | PDI | Dh (nm) | PDI | Dh (nm) | PDI | Dh (nm) | PDI |
| AmB@AB | 31.1 ± 1.7 | 0.172 ± 0.032 | 35.5 ± 2.8 | 0.189 ± 0.081 | 57.7 ± 4.3 | 0.324 ± 0.072 | 69.0 ± 6.0 | 0.234 ± 0.032 |
| AmB@AB-RET | 57.5 ± 4.2 | 0.213 ± 0.023 | 80.5 ± 7.5 | 0.204 ± 0.035 | 91.0 ± 3.5 | 0.391 ± 0.058 | 115.3 ± 7.2 | 0.359 ± 0.043 |
| AmB@ABA | 22.0 ± 2.3 | 0.198 ± 0.035 | 32.4 ± 3.3 | 0.212 ± 0.014 | 45.9 ± 4.7 | 0.429 ± 0.063 | 67.8 ± 3.5 | 0.354 ± 0.065 |
| AmB@ABA-RET | 53.4 ± 7.3 | 0.321 ± 0.041 | 77.6 ± 9.1 | 0.271 ± 0.037 | 80.2 ± 5.9 | 0.407 ± 0.024 | 81.7 ± 5.2 | 0.428 ± 0.027 |

Figure 4 shows a comparison of the distribution plots of empty ABA and ABA-RET with their corresponding formulations at 37 °C. In the presence of AmB, the diameter of the nanoparticles increased markedly, suggesting that AmB is not only dissolved in the hydrophobic micelle nucleus but is also involved in particle formations, which we assume is favoured by the hydrophobic AmB–polymer interactions. Transmission electron microscopy images provided in Figure S2, indicate that AmB loaded micelles are spherical with diameters below 100 nm.

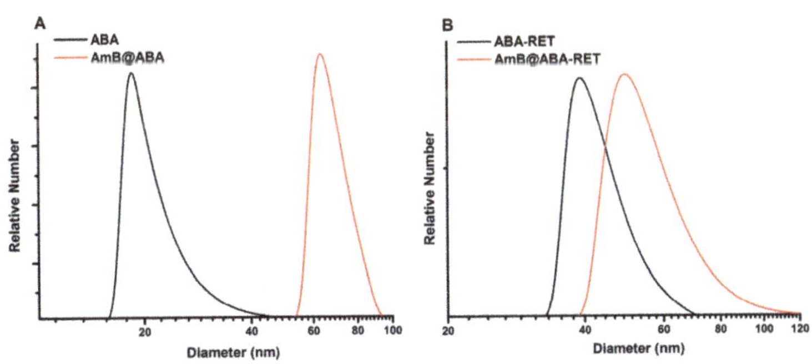


**Figure 4.** Diameter distributions of empty micelles and AmB@PMS obtained from ABA (**A**) and ABA-RET (**B**) copolymers.

### *3.5. Assessment of AmB Aggregation*

AmB, given its amphiphilic character, tends to aggregate when its concentration exceeds CAC. According to Zielińska et al. [61], the aggregation proceeds via hydrophobic interactions, while the interactions between polar groups and their amphoteric character have minimal contribution. Assessing the aggregation state of AmB in antifungal formulations allows for an estimation of its toxicity and effectivity. The toxicity of AmB is mainly adjudicated to water-soluble aggregates (including dimers), whereas insoluble poly-aggregates have been reported to be safer [62]. Monomeric AmB that accounts for a high antifungal action, presents a minimal toxicity compared with the soluble aggregates [25].

UV-vis spectroscopy is a reliable technique to characterize the aggregation state of AmB because its spectral features depend on the molecular arrangements obtained by its self-aggregation in the presence of surfactants, solvents and biomolecules. The spectrum of AmB dissolved in DMSO corresponding to the monomeric state [63] exhibited $\lambda_{max}$ at 415, 391, and 371 nm (Figure 5A). Conversely, in the spectrum of aqueous AmB, absorption maxima at 408, 385, and 365 nm are principally due to the monomeric form of the drug [64]. The absorption band at 345 nm is indicative of its aggregated form (Figure 5B). Similar tendencies are observed in the spectrum of Fungizone® (Figure 5C) and those of AmB@PMs obtained from ABA and ABA-RET (Figure 5D,E).

The AmB spectra were further processed by fitting them to Gaussian peaks as a means to estimate the relative amount of each type of arrangement. Figure 5B shows the spectrum of AmB in PBS and the corresponding fitted spectrum. In this case, five Gaussian peaks were necessary to fit the experimental data. The peaks at 327 and 346 nm correspond to different aggregates of AmB (AI and AII) with different molecular arrangements of AmB and intermolecular distances. The contributions of the monomeric form on the absorption spectra are also observed. The spectra of Fungizone® have a similar absorption given the same wavelength observed for AmB dissolved in PBS, suggesting similar arrangements of AmB in the aggregates and in the monomeric form.

As for the encapsulation of AmB in PM, the absorptions were dependent on the composition of the copolymer. In the ABA, the aggregates presented absorption at the same $\lambda_{max}$ of AmB in PBS and Fungizone®. In the presence of RET, the absorption of this fraction of AmB is shifted to a relatively high $\lambda$, suggesting an increase in distance between AmB molecules involved in the formation of aggregates, and thus less-dense structures are obtained [65]. In both formulations, the absorption at the longest wavelength corresponding to monomeric species gave rise to tow absorption bands with $\lambda_{max}$ at 423 and 408 nm. The bathochromic absorption suggests that a significant fraction of AmB underwent an increase in hydrophobicity around the heptaene moieties [22]; this finding provides evidence on the nature of the drug–polymer interaction that drive its encapsulation.

In UV–vis spectrometry, absorbance in a given $\lambda$ is proportional to the concentration of the absorbing compound. Here, the fitting peaks are integrated to let for the obtainment of the data provided in Table 5. The relative area of the peaks with relatively large $\lambda$ is proportional to the fraction of aggregated AmB. These values indicate that AmB, in its encapsulation for on PMs, is mostly in its monomeric form. By contrast, AmB dissolved in PBS and Fungizone® is mostly aggregated, and the minor fraction of monomers corresponds to CAC. Additionally, the relative amount of each type of aggregate is evaluated by determining the relative area of the aggregated AI type. The relative amounts are similar for both types of aggregates in AmB dissolved in PBS; however, the aggregates in Fungizone® with closely packed AmB molecules (AI) are the predominant arrangements. These results agree with the blue-shifting absorption associated with the aggregates previously reported for AmB in Fungizone® formulations [65,66].

In the case of micellar formulations, the aggregate distribution depends on the copolymer structures. For ABA, the aggregate type AII is the predominant arrangement, suggesting that the hydrophobic segments of PCL can reduce the formation of tight aggregates, probably by reducing the mobility of AmB molecules dissolved in the core of micelles. In the presence of RET, aggregates type AI is not detected; instead, the weak absorption at 354 nm suggests the formation of a new type of molecular arrangement that presumably involves RET moieties.

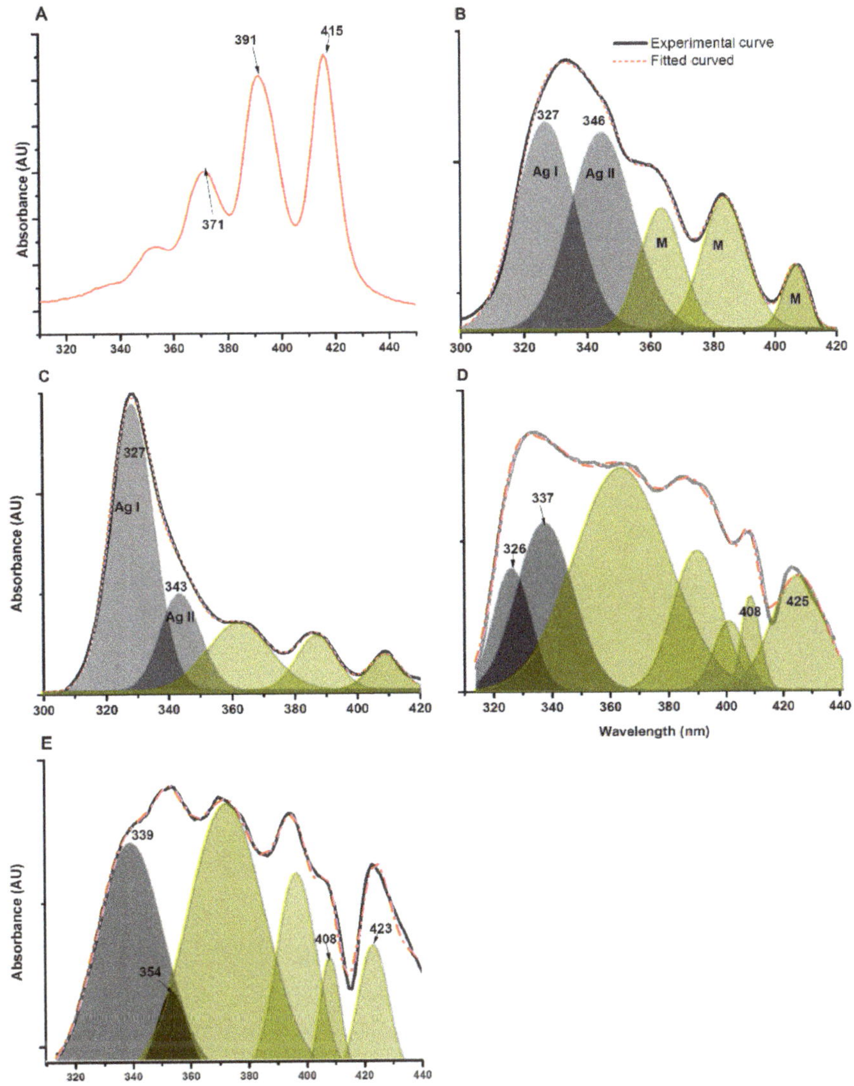


**Figure 5.** UV–vis spectrum of AmB dissolved in PBS (**A**) in water (**B**), Fungizone® (**C**), AmB@ABA (**D**) and AmB@ABA-RET (**E**). The spectra were fitted to Gaussian peaks to allow for the separation of absorption peaks due to two types of aggregates, designated as AI and AII, and the monomeric form (M).

**Table 5.** Summary of UV-vis spectrum analysis.

| Sample | Aggregation Ratio * | AgI/Ag ** |
|---|---|---|
| AmB in PBS | 0.66 | 0.49 |
| Fungizone® | 0.70 | 0.74 |
| AmB@AB | 0.21 | 0.17 |
| AmB@AB-RET | 0.32 | 0.75 |
| AmB@ABA | 0.23 | 0.14 |
| AmB@ABA-RET | 0.37 | 0.81 |

* Estimated as the ratio of the area of the absorption due to the aggregates to the total area. ** Ratio of area of the absorption of AI to the total area of the absorptions due to the aggregates.

### *3.6. Release Study*

The release of AmB from micellar formulations was assessed at 37 °C by dialysis and by using a solution of sodium deoxycholate and DMSO as the release medium to guarantee sink conditions (AmB solubility > 20 μg/mL). The concentration of AmB released to the medium was monitored using UV-vis spectroscopy at a wavelength of 406 nm. The release profiles are shown in Figure 6. The curves characterize two different behaviours. During the first 6 h, AmB is quickly released to obtain a cumulative release close to 30%. After 6 h, the release rate decreased, but the formulations continued to release the drug at a near-constant rate.

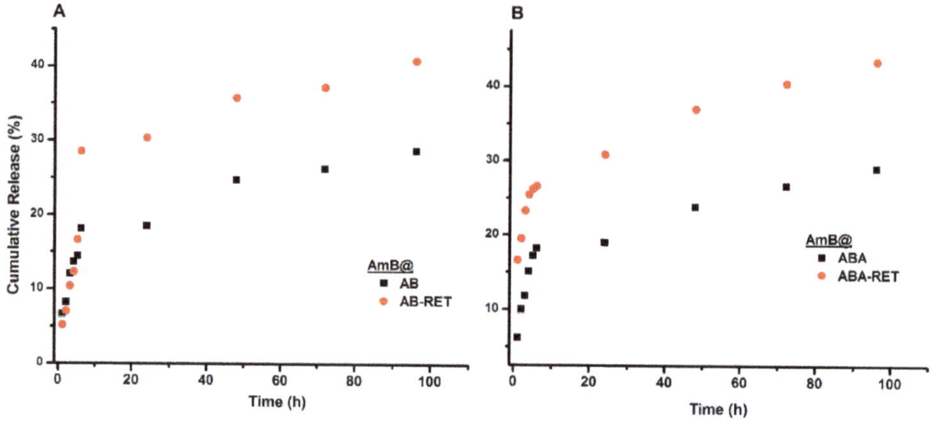


**Figure 6.** Release profiles of AmB loaded in AB (**A**) and ABA (**B**) and their corresponding conjugates with RET.

In determining the effect of RET presence on release behaviour, the profiles of representative formulations of AmB@ABA and AmB@ABA-RET were analysed on the basis of the models described in Table S1. The data corresponding to each step shown in the release profiles in Figure 6B are independently fitted to the equation representing each model. The derived values are listed in Table 6. For the initial stage (first 6 h), the release profiles fitted well with the Korsmeyer–Peppas model [67,68]. Meanwhile, AmB@ABA obtained an $n$ value of 0.68 for the supposed spherical particles, indicating an anomalous behaviour. In other words, AmB can be released by both diffusion and swelling mechanisms. The value of 0.283 obtained for AmB@ABA-RET agrees well with the diffusion or Fickian mechanism. The same trend for the second stage can be deduced from the fitting parameters. Both formulations exhibited relatively low k values that agree with the reduction of the releasing rate at prolonged times. The value of this parameter in both stages also suggest that the conjugation with RET facilities the release of AmB, presumably due to the relatively low crystallinity and thus less-dense micelle nuclei.

**Table 6.** Kinetics analysis for the release models.

| Model | Parameter | ABA | ABA-RET | ABA | ABA-RET |
|---|---|---|---|---|---|
| **Zero order** | $R^2$ | 0.975 | 0.903 | 0.973 | 0.969 |
| | K (%·$h^{-1}$) | 2.426 | 2.061 | 0.138 | 0.173 |
| **First order** | $R^2$ | 0.917 | 0.878 | 0.950 | 0.950 |
| | k ($h^{-1}$) | 0.207 | 0.095 | 0.006 | 0.005 |
| **Higuchi** | $R^2$ | 0.992 | 0.955 | 0.996 | 0.994 |
| | k (%·$h^{-0.5}$) | 0.085 | 0.074 | 0.021 | 0.026 |
| **Korsmeyer–Peppas** | $R^2$ | 0.994 | 0.975 | 0.999 | 0.999 |
| | n | 0.608 | 0.283 | 0.613 | 0.496 |
| | k ($h^{-n}$) | 0.063 | 0.166 | 0.072 | 0.141 |

In vivo, the release of AmB is facilitated by the interaction of the drug and nanocarriers with plasmatic constituents such as lipoproteins. Although in vitro those conditions cannot be emulated, a release medium that provides sink conditions allows for comparing the ability of nanostructured vehicles to control the release of AmB.

### *3.7. In Vitro Biocompatibility*

The haemolytic behaviour of AmB@PMs was compared with that of Fungizone®. RBCs were incubated with different amounts of each formulation. The haemolysis extent is plotted in Figure 7A. As previously discussed, the toxicity of AmB against mammalian cells, such as RBCs, is attributed to the formation of aggregates. This phenomenon results in the complexation of cholesterol in cell membranes and subsequently leads to the formation of pores, which then serve as leaking points of essential substances. Fungizone® is a colloidal dispersion of AmB stabilized by sodium deoxycholate (i.e., used to improve its solubility). In concordance with its high aggregation deduced from the UV-vis analysis, this formulation presented relatively high haemolytic activity, which is more pronounced at the concentrations superior to CAC of AmB (1 μmol/L). By contrast, when AmB is encapsulated in polymer micelles (i.e., to provide controlled release and avoid aggregation), its cytotoxicity is satisfactorily reduced in all concentrations. Micellar formulations with RET-conjugated precursors are more haemolytic than the corresponding AB and ABA precursors. Figure S3 shows a plot of the haemolysis extent for the micellar formulations and Fungizone® as a function of their aggregation ratios (Table 5) with concentrations of approximately 4.2 μmol/L. As the aggregation of AmB increases, their haemolytic activity also increases, indicating that reducing the aggregation of AmB lets for safe formulations.

The cytotoxicity of AmB@PMs was further studied against fibroblast L929, and compared with Fungizone®. According to Figure 7B, the micellar formulations elicited lower toxicity. At the highest concentrations of AmB of 16 μmol/L, while Fungizone®-treated cells presented a viability lower than 50%, the cytotoxicity of the micellar formulations was lower than 30%. As formerly discussed, one of the main concerns in the administration of AmB is its renal toxicity, which is related to the ability of this drug to interact with cholesterol and induce the formation of pores in mammal cell membranes. A decrease of haemolytic activity suggests that the developed PMs attenuated the intrinsic toxicity of AmB, thereby a reduction of the in vivo nephrotoxicity is expected [69]. Likewise, fibroblasts L929 proliferation assays are in concordance with biocompatible formulations [70].

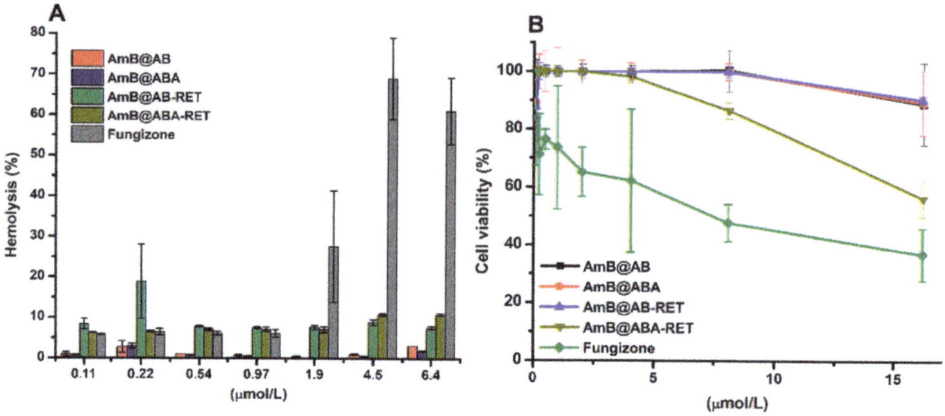


**Figure 7.** (**A**) Haemolytic profiles of AmB@PMs and Fungizone® and (**B**) Cell viability of mouse fibroblasts (L929) in the presence of AmB formulations.

### 3.8. In Vitro Effectiveness

The efficacies of the micellar formulations of AmB and Fungizone® were assessed against a set of reference ATCC and clinical strains, as listed in Table 7, along with their susceptibility profile determined according to CLSI [68]. In this group, an isolate resistant to conventional antifungals of parenteral administration was also included. In all cases, MIC was taken as the minimum concentration of AmB for the inhibited growth of yeast, the values were determined visually and confirmed employing resazurin. This violet substance is transformed to pink upon entering living cells, enabling to differentiate between alive and dead fungal cells, as shown in Figure S4 [47,48]. The derived results indicate that the broad spectrum of action of AmB can be maintained. In the case of *Candida auris* 537 resistant to AmB, the formulation of AmB@ABA inhibits its growing at 0.93 μg/mL concentration compared with the 3.75 μg/mL concentration for Fungizone®. Moreover, this result was further confirmed by the growing plots depicted in Figure S5. In most cases, micellar formulations exhibit higher efficacy than Fungizone® against the evaluated strains, as deduced from the measured MIC values.

**Table 7.** Values of minimum inhibitory concentration (MIC) of AmB@PMs formulations compare with Fungizone® and profiles of susceptibilities of *Candida spp.* to fluconazol and caspofungine by CLSI reference methodology.

| Yeast | Reference | CLSI | | Formulations | | | | |
|---|---|---|---|---|---|---|---|---|
| | | FLC * | CAS * | Fungizone® | AmB@AB | AmB@ABA | AmB@AB-RET | AmB@ABA-RET |
| | | MIC (μg/mL) | | | | | | |
| *C. albicans* | SC5314 | 1 | 0.06 | 0.46 | <0.11 | <0.11 | <0.11 | <0.11 |
| *C. glabrata* | ATCC 2001 | 0.25 | 0.06 | 0.46 | <0.11 | <0.11 | <0.11 | <0.11 |
| *C. krusei* | ATCC 6258 | 32 | 0.25 | 1.875 | 0.23 | 0.23 | 0.43 | 0.23 |
| *C. parapsilosis* | ATCC 22019 | 1 | 0.25 | 0.23 | <0.11 | <0.11 | <0.11 | <0.11 |
| *C. auris* | 435-PUJ-HUSI | 8 | 0.25 | 0.93 | 0.23 | 0.23 | 0.46 | 0.23 |
| *C. auris* * | 537-PUJ-HUSI | 128 ** | <0.05 | 3.75 | 1.865 | 0.93 | 3.75 | 3.75 |
| *C. parapsilosis* | 75-PUJ-FVL | 0.125 | <0.05 | 0.11 | <0.11 | <0.11 | 0.11 | <0.11 |

* FLC: Fluconazole*CAS: caspofungin. ** This MIC corresponds to a resistance isolate.

## 4. Conclusions

Copolymers composed of PEG and PCL are promising as micellar vehicles for drug delivery because of their biocompatibility. However, the absence of functional groups on PCL hinders their interaction with drugs which in turn results in low loading capacities. It was demonstrated that upon the conjugation of these block copolymers with RET, the encapsulation of AmB was enhanced to achieve values comparable with commercial liposomal formulations such as AmBisome, Lambin and Lipholyn and superior to Fungisome [12]. Upon its encapsulation, AmB retained its antifungal spectrum and presented an attenuated toxicity. Moreover, these nanocarriers improved the antifungal activity of AmB against most of the tested yeast strains as deduced from a reduction of the MICs of up to eight-fold compared with Fungizone®. Both, a high loading capacity and an improved efficacy are of paramount importance for the translation of new formulations to clinical practice, because they allow for more favourable cost-benefit ratios, and also a reduction of the toxicity risks associated to the polymeric vehicles and their metabolic products.

**Supplementary Materials:** The following are available online at http://www.mdpi.com/1999-4923/12/3/196/s1, Figure S1. 1H-NMR spectra of A. ABA and B. ABA-Succinic acid copolymers. Figure S2. TEM image of micellar dispersion of sample AmB@ABA-RET showing the presence of spherical nanoparticles. Figure S3. Plot of relative haemolysis of each formulation at the concentration of 3.8 μg/mL versus its corresponding aggregation ratio value listed in Table 5, Figure S4. Determination of MIC employing resazurin. Figure S5. Growing curves of *C. auris* 537 in the presence of different concentrations of AmB@PMs, Table S1. Release models.

**Author Contributions:** Conceptualization, L.D.P. and C.M.P.-G.; methodology, Y.J.R., L.F.Q. and J.C.V.; software, L.F.Q.; validation, Y.J.R., L.F.Q. and J.C.V.; formal analysis, L.D.P., Y.B. and C.M.P.-G.; investigation, L.D.P., Y.B. and C.M.P.-G.; resources, L.D.P., Y.B. and C.M.P.-G.; data curation, L.D.P., Y.B. and C.M.P.-G.; writing—original draft preparation, L.D.P.; writing—review and editing, L.D.P., Y.B. and C.M.P.-G.; visualization, C.M.P.-G.; supervision, L.D.P., Y.B. and C.M.P.-G.; project administration, L.D.P.; funding acquisition, L.D.P., Y.B. and C.M.P.-G. All authors have read and agreed to the published version of the manuscript.

**Funding:** This work was supported by Colciencias–Colombia through the Grant number: 834-2017.

**Conflicts of Interest:** The authors declare no conflict of interest.

## References

1. Sifuentes-Osornio, J.; Corzo-León, D.E.; Ponce-De-León, L.A. Epidemiology of invasive fungal infections in Latin America. *Curr. Fungal Infec. Rep.* **2012**, *6*, 23–34. [CrossRef]
2. Ascioglu, S.; Rex, J.; De Pauw, B.; Bennett, J.; Bille, J.; Crokaert, F.; Denning, D.; Donnelly, J.; Edwards, J.; Erjavec, Z. Defining opportunistic invasive fungal infections in immunocompromised patients with cancer and hematopoietic stem cell transplants: An international consensus. *Clin. Infect. Dis.* **2002**, *34*, 7–14. [CrossRef] [PubMed]
3. Mirsaeidi, M.; Motahari, H.; Taghizadeh Khamesi, M.; Sharifi, A.; Campos, M.; Schraufnagel, D.E. Climate Change and Respiratory Infections. *Ann. Am. Thorac. Soc.* **2016**, *13*, 1223–1230. [CrossRef] [PubMed]
4. Chen, K.; Wang, Q.; Pleasants, R.A.; Ge, L.; Liu, W.; Peng, K.; Zhai, S. Empiric treatment against invasive fungal diseases in febrile neutropenic patients: A systematic review and network meta-analysis. *BMC Infect. Dis.* **2017**, *17*, 159. [CrossRef] [PubMed]
5. Saravolatz, L.D.; Ostrosky-Zeichner, L.; Marr, K.A.; Rex, J.H.; Cohen, S.H. Amphotericin B: Time for a New "Gold Standard". *Clin. Infect. Dis.* **2003**, *37*, 415–425. [CrossRef] [PubMed]
6. Sarosi, G.A.; Amphotericin, B. Still the 'gold standard' for antifungal therapy. *Postgrad Med.* **1990**, *88*, 151–152, 155–161, 165–166. [CrossRef]
7. Calvo, B.; Melo, A.S.A.; Perozo-Mena, A.; Hernandez, M.; Francisco, E.C.; Hagen, F.; Meis, J.F.; Colombo, A.L. First report of Candida auris in America: Clinical and microbiological aspects of 18 episodes of candidemia. *J. Infect.* **2016**, *73*, 369–374. [CrossRef]
8. Vogelsinger, H.; Weiler, S.; Djanani, A.; Kountchev, J.; Bellmann-Weiler, R.; Wiedermann, C.J.; Bellmann, R. Amphotericin B tissue distribution in autopsy material after treatment with liposomal amphotericin B and amphotericin B colloidal dispersion. *J. Antimicrob. Chemother.* **2006**, *57*, 1153–1160. [CrossRef]
9. Kamiński, D.M. Recent progress in the study of the interactions of amphotericin B with cholesterol and ergosterol in lipid environments. *Eur. Biophys. J.* **2014**, *43*, 453–467. [CrossRef]
10. Wingard, J.R.; White, M.H.; Anaissie, E.; Raffalli, J.; Goodman, J.; Arrieta, A. A randomized, double-blind comparative trial evaluating the safety of liposomal amphotericin B versus amphotericin B lipid complex in the empirical treatment of febrile neutropenia. *Clin. Infect. Dis.* **2000**, *31*, 1155–1163. [CrossRef]

11. Bates, D.; Su, L.; Yu, D.; Chertow, G.; Seger, D.; Gomes, D.; Dasbach, E.; Platt, R. Mortality and costs of acute renal failure associated with amphotericin B therapy. *Clin. Infect. Dis.* **2001**, *32*, 686–693. [CrossRef] [PubMed]
12. Adler-Moore, J.P.; Gangneux, J.-P.; Pappas, P.G. Comparison between liposomal formulations of amphotericin B. *Med. Mycol.* **2016**, *54*, 223–231. [CrossRef] [PubMed]
13. Fanos, V.; Cataldi, L. Amphotericin B-induced nephrotoxicity: A review. *J. Chemother.* **2000**, *12*, 463–470. [CrossRef] [PubMed]
14. Ringdén, O.; Meunier, F.; Tollemar, J.; Ricci, P.; Tura, S.; Kuse, E.; Viviani, M.A.; Gorin, N.C.; Klastersky, J.; Fenaux, P.; et al. Efficacy of amphotericin B encapsulated in liposomes (AmBisome) in the treatment of invasive fungal infections in immunocompromised patients. *J. Antimicrob. Chemother.* **1991**, *28* (Suppl. B), 73–82.
15. Walsh, T.J.; Finberg, R.W.; Arndt, C.; Hiemenz, J.; Schwartz, C.; Bodensteiner, D.; Pappas, P.; Seibel, N.; Greenberg, R.N.; Dummer, S.; et al. Liposomal Amphotericin B for Empirical Therapy in Patients with Persistent Fever and Neutropenia. *N. Engl. J. Med.* **1999**, *340*, 764–771. [CrossRef]
16. Botero Aguirre, J.P.; Restrepo Hamid, A.M. Amphotericin B deoxycholate versus liposomal amphotericin B: Effects on kidney function. *Cochrane Database Syst. Rev.* **2015**, *11*. [CrossRef]
17. Gamboa Garay, O.A.; Fuentes Pachón, J.C.; Cuervo Maldonado, S.I.; Gómez Rincón, J.C.; Castillo Londoño, J.S. Análisis de Costo Efectividad de Estrategias de Tratamiento Antimicótico en Pacientes con Neutropenia Febril Persistente y Tratamiento Antibiótico de Amplio Espectro. *Value Health Reg. Issues* **2012**, *1*, 201–210. [CrossRef]
18. Jensen, G.; Skenes, C.; Bunch, T.; Weissman, C.; Amirghahari, N.; Satorius, A.; Moynihan, K.; Eley, C. Determination of the relative toxicity of amphotericin B formulations: A red blood cell potassium release assay. *Drug Deliv.* **1999**, *6*, 81–88. [CrossRef]
19. Perlin, D.S.; Rautemaa-Richardson, R.; Alastruey-Izquierdo, A. The global problem of antifungal resistance: Prevalence, mechanisms, and management. *Lancet Infect. Dis.* **2017**, *17*, e383–e392. [CrossRef]
20. Fernández-García, R.; de Pablo, E.; Ballesteros, M.P.; Serrano, D.R. Unmet clinical needs in the treatment of systemic fungal infections: The role of amphotericin B and drug targeting. *Int. J. Pharm.* **2017**, *525*, 139–148. [CrossRef] [PubMed]
21. Kasai, Y.; Nobuaki, M.; Hiroyuki, U.; Kenichi, N.; Shinya, Y.; Murata, M.; Tohru, O. Synthesis of 6-F-Ergosterol and Its Influence on Membrane-Permeabilization of Amphotericin B and Amphidinol 3. *Org. Biomol. Chem.* **2011**, *9*, 1437–1442. [CrossRef] [PubMed]
22. Mouri, R.; Konoki, K.; Matsumori, N.; Oishi, T.; Murata, M. Complex formation of amphotericin B in sterol-containing membranes as evidenced by surface plasmon resonance. *Biochemistry* **2008**, *47*, 7807–7815. [CrossRef] [PubMed]
23. Nakagawa, Y.; Umegawa, Y.; Nonomura, K.; Matsushita, N.; Takano, T.; Tsuchikawa, H.; Hanashima, S.; Oishi, T.; Matsumori, N.; Murata, M. Axial hydrogen at C7 position and bumpy tetracyclic core markedly reduce sterol's affinity to amphotericin B in membrane. *Biochemistry* **2015**, *54*, 303–312. [CrossRef]
24. Barwicz, J.; Tancrède, P. The effect of aggregation state of amphotericin-B on its interactions with cholesterol- or ergosterol-containing phosphatidylcholine monolayers. *Chem. Phys. Lipids* **1997**, *85*, 145–155. [CrossRef]
25. Neumann, A.; Baginski, M.; Czub, J. Exploring Amphotericin B-Membrane Interactions: Free Energy Simulations. *Biophys. J.* **2013**, *104* (Suppl. 1), 250. [CrossRef]
26. Legrand, P.; Romero, E.A.; Cohen, B.E.; Bolard, J. Effects of aggregation and solvent on the toxicity of amphotericin B to human erythrocytes. *Antimicrob. Agents Chemother.* **1992**, *36*, 2518–2522. [CrossRef] [PubMed]
27. Blanco, E.; Shen, H.; Ferrari, M. Principles of nanoparticle design for overcoming biological barriers to drug delivery. *Nat. Biotechnol.* **2015**, *33*, 941. [CrossRef]
28. Cabral, H.; Matsumoto, Y.; Mizuno, K.; Chen, Q.; Murakami, M.; Kimura, M.; Terada, Y.; Kano, M.; Miyazono, K.; Uesaka, M. Accumulation of sub-100 nm polymeric micelles in poorly permeable tumours depends on size. *Nat. Nanotechnol.* **2011**, *6*, 815. [CrossRef]
29. Lavasanifar, A.; Samuel, J.; Sattari, S.; Kwon, G.S. Block copolymer micelles for the encapsulation and delivery of amphotericin B. *Pharm. Res.* **2002**, *19*, 418–422. [CrossRef]
30. Kataoka, K.; Harada, A.; Nagasaki, Y. Block copolymer micelles for drug delivery: Design, characterization and biological significance. *Adv. Drug Deliv. Rev.* **2012**, *64*, 37–48. [CrossRef]

31. Sun, X.; Wang, G.; Zhang, H.; Hu, S.; Liu, X.; Tang, J.; Shen, Y. The Blood Clearance Kinetics and Pathway of Polymeric Micelles in Cancer Drug Delivery. *ACS Nano* **2018**, *12*, 6179–6192. [CrossRef] [PubMed]
32. Han, K.; Miah, J.; Shanmugam, S.; Yong, C.S.; Choi, H.-G.; Kim, J.A.; Yoo, B.K. Mixed micellar nanoparticle of amphotericin B and poly styrene-block-poly ethylene oxide reduces nephrotoxicity but retains antifungal activity. *Arch. Pharmacal Res.* **2007**, *30*, 1344–1349.
33. Xu, H.; Teng, F.; Zhou, F.; Zhu, L.; Wen, Y.; Feng, R.; Song, Z. Linolenic acid-modified MPEG-PEI micelles for encapsulation of amphotericin B. *Future Med. Chem.* **2019**, *11*, 2647–2662. [CrossRef] [PubMed]
34. Diezi, T.A.; Takemoto, J.K.; Davies, N.M.; Kwon, G.S. Pharmacokinetics and nephrotoxicity of amphotericin B-incorporated poly(ethylene glycol)-block-poly(N-hexyl stearate l-aspartamide) micelles. *J. Pharm. Sci.* **2011**, *100*, 2064–2070. [CrossRef]
35. Yoo, B.K.; Jalil Miah, M.A.; Lee, E.S.; Han, K. Reduced renal toxicity of nanoparticular amphotericin B micelles prepared with partially benzylated poly-L-aspartic acid. *Biol. Pharm. Bull.* **2006**, *29*, 1700–1705. [CrossRef]
36. Alconcel, S.N.; Baas, A.S.; Maynard, H.D. FDA-approved poly (ethylene glycol)–protein conjugate drugs. *Polym. Chem.* **2011**, *2*, 1442–1448. [CrossRef]
37. Williford, J.-M.; Archang, M.M.; Minn, I.; Ren, Y.; Wo, M.; Vandermark, J.; Fisher, P.B.; Pomper, M.G.; Mao, H.-Q. Critical length of PEG grafts on lPEI/DNA nanoparticles for efficient in vivo delivery. *ACS Biomater. Sci. Eng.* **2016**, *2*, 567–578. [CrossRef]
38. Pozzi, D.; Colapicchioni, V.; Caracciolo, G.; Piovesana, S.; Capriotti, A.L.; Palchetti, S.; De Grossi, S.; Riccioli, A.; Amenitsch, H.; Laganà, A. Effect of polyethyleneglycol (PEG) chain length on the bio–nano-interactions between PEGylated lipid nanoparticles and biological fluids: From nanostructure to uptake in cancer cells. *Nanoscale* **2014**, *6*, 2782–2792. [CrossRef]
39. Yoon, K.; Kang, H.C.; Li, L.; Cho, H.; Park, M.-K.; Lee, E.; Bae, Y.H.; Huh, K.M. Amphiphilic poly (ethylene glycol)-poly (ε-caprolactone) AB 2 miktoarm copolymers for self-assembled nanocarrier systems: Synthesis, characterization, and effects of morphology on antitumor activity. *Polym. Chem.* **2015**, *6*, 531–542. [CrossRef]
40. Castleberry, S.A.; Quadir, M.A.; Sharkh, M.A.; Shopsowitz, K.E.; Hammond, P.T. Polymer conjugated retinoids for controlled transdermal delivery. *J. Control. Release* **2017**, *262*, 1–9. [CrossRef]
41. Maiti, S.; Chatterji, P.R.; Nisha, C.; Manorama, S.; Aswal, V.K.; Goyal, P.S. Aggregation and polymerization of PEG-based macromonomers with methacryloyl group as the only hydrophobic segment. *J. Colloid Interface Sci.* **2001**, *240*, 630–635. [CrossRef]
42. Shim, W.S.; Kim, S.W.; Choi, E.K.; Park, H.J.; Kim, J.S.; Lee, D.S. Novel pH sensitive block copolymer micelles for solvent free drug loading. *Macromol. Biosci.* **2006**, *6*, 179–186. [CrossRef]
43. Diaz, I.L.; Perez, L.D. Synthesis and micellization properties of triblock copolymers PDMAEMA-b-PCL-b-PDMAEMA and their applications in the fabrication of amphotericin B-loaded nanocontainers. *Colloid Polym. Sci.* **2015**, *293*, 913–923. [CrossRef]
44. Villamil, J.C.; Parra-Giraldo, C.M.; Pérez, L.D. Enhancing the performance of PEG-b-PCL copolymers as precursors of micellar vehicles for amphotericin B through its conjugation with cholesterol. *Colloids Surf. A Physicochem. Eng. Asp.* **2019**, *572*, 79–87. [CrossRef]
45. Ramage, G.; Walle, K.V.; Wickes, B.L.; López-Ribot, J.L. Standardized method for in vitro antifungal susceptibility testing of Candida albicansbiofilms. *Antimicrob. Agents Chemother.* **2001**, *45*, 2475–2479. [CrossRef]
46. Lockhart, S.R.; Bolden, C.B.; Iqbal, N.; Kuykendall, R.J. Validation of 24-hour flucytosine MIC determination by comparison with 48-hour determination by the Clinical and Laboratory Standards Institute M27-A3 broth microdilution reference method. *J. Clin. Microbiol.* **2011**, *49*, 4322–4325. [CrossRef]
47. Fai, P.B.; Grant, A. A rapid resazurin bioassay for assessing the toxicity of fungicides. *Chemosphere* **2009**, *74*, 1165–1170. [CrossRef]
48. O'brien, J.; Wilson, I.; Orton, T.; Pognan, F. Investigation of the Alamar Blue (resazurin) fluorescent dye for the assessment of mammalian cell cytotoxicity. *Eur. J. Biochem.* **2000**, *267*, 5421–5426. [CrossRef]
49. Chen, Y.J.; Fang, H.J.; Hsu, S.C.N.; Jheng, N.Y.; Chang, H.C.; Ou, S.W.; Peng, W.T.; Lai, Y.C.; Chen, J.Y.; Chen, P.L.; et al. Improving the ring-opening polymerization of ε-caprolactone and l-lactide using stannous octanoate. *Polym. Bull.* **2013**, *70*, 993–1001. [CrossRef]

50. Erothu, H.; Sohdi, A.A.; Kumar, A.C.; Sutherland, A.J.; Dagron-Lartigau, C.; Allal, A.; Hiorns, R.C.; Topham, P.D. Facile synthesis of poly (3-hexylthiophene)-block-poly (ethylene oxide) copolymers via Steglich esterification. *Polym. Chem.* **2013**, *4*, 3652–3655. [CrossRef]
51. Liu, J.; Zeng, F.; Allen, C. In vivo fate of unimers and micelles of a poly(ethylene glycol)-block-poly(caprolactone) copolymer in mice following intravenous administration. *Eur. J. Pharm. Biopharm.* **2007**, *65*, 309–319. [CrossRef]
52. Kim, S.; Shi, Y.; Kim, J.Y.; Park, K.; Cheng, J.-X. Overcoming the barriers in micellar drug delivery: Loading efficiency, in vivo stability, and micelle—cell interaction. *Expert Opin. Drug Deliv.* **2010**, *7*, 49–62. [CrossRef]
53. Ohyashiki, T.; Mohri, T. Fluorometric Analysis of the Micelle Formation Process of Surfactants in Aqueous Solution. I. Utility of Pyrene in Determination of the Critical Micelle Concentration. *Chem. Pharm. Bull.* **1983**, *31*, 1296–1300. [CrossRef]
54. Scholz, N.; Behnke, T.; Resch-Genger, U. Determination of the Critical Micelle Concentration of Neutral and Ionic Surfactants with Fluorometry, Conductometry, and Surface Tension—A Method Comparison. *J. Fluoresc.* **2018**, *28*, 465–476. [CrossRef]
55. Bogdanov, B.; Vidts, A.; Van Den Buicke, A.; Verbeeck, R.; Schacht, E. Synthesis and thermal properties of poly(ethylene glycol)-poly(ϵ-caprolactone) copolymers. *Polymer* **1998**, *39*, 1631–1636. [CrossRef]
56. Huang, Y.; Li, L.; Li, G. An enzyme-catalysed access to amphiphilic triblock copolymer of PCL-b-PEG-b-PCL: Synthesis, characterization and self-assembly properties. *Des. Monomers Polym.* **2015**, *18*, 799–806. [CrossRef]
57. Verheyen, S.; Augustijns, P.; Kinget, R.; Van den Mooter, G. Melting behavior of pure polyethylene glycol 6000 and polyethylene glycol 6000 in solid dispersions containing diazepam or temazepam: A DSC study. *Thermochim. Acta* **2001**, *380*, 153–164. [CrossRef]
58. Liu, X.-Q.; Bao, R.-Y.; Wu, X.-J.; Yang, W.; Xie, B.-H.; Yang, M.-B. Temperature induced gelation transition of a fumed silica/PEG shear thickening fluid. *RSC Adv.* **2015**, *5*, 18367–18374. [CrossRef]
59. Raman, A.S.; Vishnyakov, A.; Chiew, Y.C. A coarse-grained model for PCL: Conformation, self-assembly of MePEG-b-PCL amphiphilic diblock copolymers. *Mol. Simul.* **2017**, *43*, 92–101. [CrossRef]
60. Linse, P.; Malmsten, M. Temperature-dependent micellization in aqueous block copolymer solutions. *Macromolecules* **1992**, *25*, 5434–5439. [CrossRef]
61. Zielińska, J.; Wieczór, M.; Bączek, T.; Gruszecki, M.; Czub, J. Thermodynamics and kinetics of amphotericin B self-association in aqueous solution characterized in molecular detail. *Sci. Rep.* **2016**, *6*, 19109. [CrossRef]
62. Espada, R.; Valdespina, S.; Alfonso, C.; Rivas, G.; Ballesteros, M.P.; Torrado, J.J. Effect of aggregation state on the toxicity of different amphotericin B preparations. *Int. J. Pharm.* **2008**, *361*, 64–69. [CrossRef]
63. Torrado, J.J.; Espada, R.; Ballesteros, M.P.; Torrado-Santiago, S. Amphotericin B formulations and drug targeting. *J. Pharm. Sci.* **2008**, *97*, 2405–2425. [CrossRef]
64. Fujii, G.; Chang, J.-E.; Coley, T.; Steere, B. The formation of amphotericin B ion channels in lipid bilayers. *Biochemistry* **1997**, *36*, 4959–4968. [CrossRef]
65. Barwicz, J.; Gruszecki, W.I.; Gruda, I. Spontaneous Organization of Amphotericin B in Aqueous Medium. *J. Colloid Interface Sci.* **1993**, *158*, 71–76. [CrossRef]
66. Stoodley, R.; Wasan, K.M.; Bizzotto, D. Fluorescence of Amphotericin B-Deoxycholate (Fungizone) Monomers and Aggregates and the Effect of Heat-Treatment. *Langmuir* **2007**, *23*, 8718–8725. [CrossRef]
67. Son, G.-H.; Lee, B.-J.; Cho, C.-W. Mechanisms of drug release from advanced drug formulations such as polymeric-based drug-delivery systems and lipid nanoparticles. *J. Pharm. Investig.* **2017**, *47*, 287–296. [CrossRef]
68. De Oliveira, E.F.; Paula, H.C.; de Paula, R.C. Alginate/cashew gum nanoparticles for essential oil encapsulation. *Colloids Surfaces B Biointerfaces* **2014**, *113*, 146–151. [CrossRef]
69. Italia, J.L.; Yahya, M.M.; Singh, D.; Ravi Kumar, M.N.V. Biodegradable nanoparticles improve oral bioavailability of amphotericin B and show reduced nephrotoxicity compared to intravenous fungizone. *Pharm. Res.* **2009**, *26*, 1324–1331. [CrossRef]
70. Jung, O.; Smeets, R.; Porchetta, D.; Kopp, A.; Ptock, C.; Müller, U.; Heiland, M.; Schwade, M.; Behr, B.; Kröger, N.; et al. Optimized in vitro procedure for assessing the cytocompatibility of magnesium-based biomaterials. *Acta Biomater.* **2015**, *23*, 354–363. [CrossRef]

MDPI
St. Alban-Anlage 66
4052 Basel
Switzerland
Tel. +41 61 683 77 34
Fax +41 61 302 89 18
www.mdpi.com

*Pharmaceutics* Editorial Office
E-mail: pharmaceutics@mdpi.com
www.mdpi.com/journal/pharmaceutics

www.ingramcontent.com/pod-product-compliance
Lightning Source LLC
Chambersburg PA
CBHW061625210926
51850CB00009B/2442
* 9 7 8 3 0 3 9 3 6 3 0 6 3 *